DESIGNING FOR PRODUCT SOUND QUALITY

T0321135

MECHANICAL ENGINEERING
A Series of Textbooks and Reference Books

Founding Editor

L. L. Faulkner

*Columbus Division, Battelle Memorial Institute
and Department of Mechanical Engineering
The Ohio State University
Columbus, Ohio*

Additional Volumes in Preparation

Mechanical Engineering Software

DESIGNING FOR PRODUCT SOUND QUALITY

RICHARD H. LYON

RH Lyon Corp
Cambridge, Massachusetts

CRC Press
Taylor & Francis Group
Boca Raton London New York

CRC Press is an imprint of the
Taylor & Francis Group, an **informa** business

CRC Press
Taylor & Francis Group
6000 Broken Sound Parkway NW, Suite 300
Boca Raton, FL 33487-2742

First issued in paperback 2019

© 2000 by Taylor & Francis Group, LLC
CRC Press is an imprint of Taylor & Francis Group, an Informa business

No claim to original U.S. Government works

ISBN-13: 978-0-8247-0400-1 (hbk)
ISBN-13: 978-0-367-39854-5 (pbk)

**Visit the Taylor & Francis Web site at
http://www.taylorandfrancis.com**

**and the CRC Press Web site at
http://www.crcpress.com**

Library of Congress Cataloging-in-Publication Data

Lyon, Richard H.
    Designing for product sound quality / Richard H. Lyon.
      p. cm. — (Mechanical engineering ; 127)
    Includes bibliographical references and index.
    ISBN 0-8247-0400-2 (alk. paper)
    1. Design, Industrial. 2. Acoustical engineering. 3. Noise control. I. Title. II.
Mechanical engineering (Marcel Dekker, Inc.) ; 127.

TS171.4. L96 2000
745.2—dc21   00-037689

PREFACE

Presenting a book on product design related to sound is a tricky business. The concern for issues in this field is a very recent affair for most products, so sound is "the new guy on the block" trying to muscle in on territory already owned by other areas: lubrication; structural integrity; heat management; and features associated with industrial design: shape, size, accessibility, and texture. After working so hard to achieve a harmonious balance (compromise) between these concerns, the design team does not look forward to contending with new considerations of sound later in the process.

Another issue is the traditional approach that the acoustics community has taken toward product sound issues—as "product noise," to be solved by "noise control." This approach treats sound as always being objectionable and to be solved by add-on treatments (structural damping, enclosures, isolators, or mufflers) after the product has otherwise been designed. Industry has thus viewed acoustics as expensive, reducing the bottom line, and limiting the domain of acoustical consultants, in favor of their own design teams.

Because of the non-involvement of acoustical specialists in product design, the rules of thumb and design charts that exist in other areas (e.g., lubrication) have not been available. One of the purposes of this book is to show designers that certain useful generalizations can be made about product sound sources and structural radiation. Even if a detailed analysis cannot be made based on this material, an understanding of the mechanisms of how sound is generated will be helpful.

This brings us to the trickiest part of the whole process. The ways in which mechanisms generate noise energy, how structures respond to that excitation, how they radiate a part of the energy as sound, and how a listener will judge the sound are among the most complex topics in acoustics. Yet the designer to whom this book is directed is supposed to understand these processes well enough to make decisions about components, structures, and assembly to achieve a sound that is acceptable to a customer!

This is fundamentally not a book about sound, vibration, or psychoacoustics; such books that I recommend are listed in a bibliography. In addition, a number of courses and seminars (some on the Web) are available, which the author highly recommends. This book is about design: how to use basic information about these topics in the process of designing a product. Just as a designer can go to a handbook to design the product for lubrication or heat transfer without being an expert in those areas, this book presents information about acoustics that the designer can use.

But, because acoustics is the new guy on the block, the development of rules of thumb, formulas, and charts for design still has a way to go, and each product design that provides new information can add to the knowledge base, both within a particular company and within an industry. Designers who make such advances are encouraged to make new information available at trade and technical meetings and in associated publications. Then we will all continue to learn.

Richard H. Lyon

CONTENTS

CHAPTER 3

NOISE CONTROL, NOISE REDUCTION, AND DESIGN FOR SOUND QUALITY   35

CHAPTER 4

SOURCES OF PRESCRIBED MOTION: IMBALANCE, CAMS, GEARS, CHAINS, AND SPROCKETS   43

CHAPTER 5

SOURCES OF PRESCRIBED FORCE: ELECTROMAGNETICS, AIRFLOW, COMBUSTION, AND IMPACT   65

CHAPTER 8

BASICS OF SOUND RADIATION 133

CHAPTER 9

SOUND RADIATION BY PRODUCTS AND ITS MEASUREMENT 153

CHAPTER 10

QUALITY ASSURANCE USING VIBRATION 173

CHAPTER **1**   **The Meaning
of Product
Sound Quality**

This chapter discusses the role of sound in product acceptance and how companies can respond in order to maintain and improve the acceptance of their products. The character of sound that relates to acceptance is called Sound Quality (SQ). The chapter indicates the role of various functions within the company in dealing with the issue of improving SQ.

1.1
PRODUCT ATTRIBUTES AND THE SPECIAL ROLE OF SOUND

All products have attributes that affect their acceptability or desirability for the user. As indicated in Figure 1.1, these attributes fall in the three general categories of Resource Commitment, Functionality, and Aesthetics. Within each category, there is a set of attributes or features that may or may not be of importance for a particular product.

    Initial cost, maintenance, space allocation, and training costs all represent aspects of Resource Commitment. If the product is a diamond ring or work of art, initial cost is an important attribute, whereas the others are not. If the product is a table saw or an automobile, all of these attributes will likely be significant. For many products,

Resource Commitment is a dominant factor in the initial purchasing decision, but once the product is purchased, other attributes may well assume a larger role in determining satisfaction.

Attributes for Functionality include productivity (What does the product do for me?), safety, reliability, and (user) friendliness. Again, all of these attributes would be important for a food processor or a space heater, but safety would not be a major issue in the case of a refrigerator or a personal computer. However, the other three attributes would be important.

Aesthetic attributes include color, form, tactility, texture, flavor, and odor. Some of these are more important for "hard goods," while others are more important for foods and perfumed goods. These attributes also include product sound, which is important for many products. But sound is also shown as an attribute for functionality because users make judgments about both function and aesthetics based on the sound. An engine sound may be pleasing or harsh to listen to, but users also make judgments about how the engine is working based on its sound.

Figure 1.1   Product attributes that determine the acceptability of products. Sound has a dual role in aesthetic quality and judgment of functionality. A vertical scale, the value of the attribute for a given product is not shown, and may well change with time.

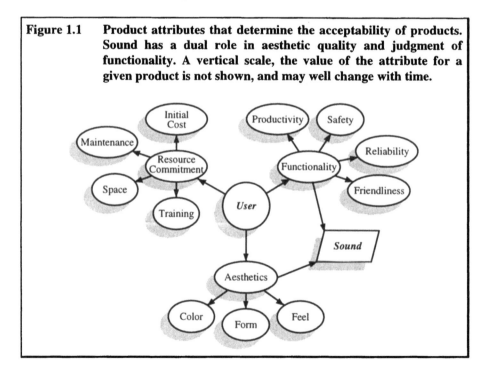

The value of a particular attribute may well change over time (the need for training in using a PC for example). This is particularly true for sound, which has become a much more important attribute for products in recent years, partly as a result of automotive industry advertising. If a good sound indicates high quality of an automobile, then it probably indicates quality in other products as well. Such a "spillover" effect happens frequently as producers follow trends that seem to work in one product area, and they then apply it to their own.

It is sometimes claimed that some attributes or product characteristics are "subjective" and therefore "merely a matter of opinion." The sound of a product is a case in point, and of course, there is a subjective aspect to sound. But all attributes, including sound, have both objective and subjective aspects. The price of a product is objective—$199.95 is an objective and quantifiable amount of money—but the subjective character of price is revealed by the fact that the price is not presented as $200.

The science of relating how people react to sounds to the physical stimulus (pressure fluctuations in the air)—is called psychoacoustics. The perceptual reaction to the sound is internal to the subject, and it is the work of psychoacoustics to first relate the perception to an objective internal scale and then to external objective physical metrics of sound. For example, the perception of loudness has been shown to be quantifiable internally (people are able to judge when one sound is twice as loud as another) and externally as a physical metric of loudness that correlates to the internal response.

1.2
HOW PRODUCING COMPANIES DEAL WITH SOUND

Sound as a product attribute is handled in different ways in the functional divisions of a product company. The "Benzene ring" model of a product company shown in Figure 1.2 breaks down these functions into the three major categories of Define, Design, and Build. For example, the Marketing function looks into the market to see how customers are responding to the products of the company and its competitors. It does this through Focus Groups; mail surveys; comments from dealers and service representatives; and customer calls, letters, and e-mail.

Figure 1.2   **This "benzene ring" model of a product company shows the main functional parts of the company as a "Define-Design-Build" triptych. Each of these has an important role in a Sound Quality program in a company.**

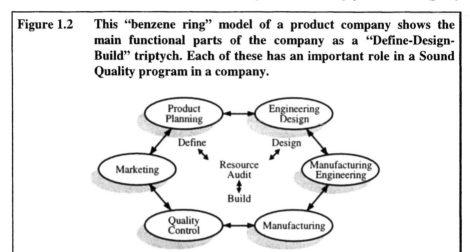

When discussing the sound attributes of their products, companies generally talk about how quiet they are. This discussion varies greatly within each industry: A "silent" computer is said to be inaudible and therefore without annoyance due to the sound. A quiet oil burner and a quiet stepping motor are said to have advanced design and are therefore quiet. A quiet ceiling fan is associated with the comfort and security of a home. Dishwashers are quiet and therefore do not interfere with other family activities. The quietness of an automobile reflects the high-tech nature of its design and thus satisfies the sophistication of both its engineers and its customers. In another automobile the sound supports and projects an image of the sports roadsters of the 50s. And a particular sound is so associated with a motorcycle that its identity is projected by its sound. These examples illustrate that product sound is not merely a negative attribute to be reduced or eliminated—it is an attribute that can be exploited for positive product presentation.

In some cases SQ can have a significant effect on market penetration. Some segments of a market may be much more sensitive to sound than others. For example, an assisted breathing medical apparatus may have potential usage in industrial plants, in hospitals, and in the home. These markets are likely to be very different in their concern about the sound, and a reduction or other modification of the sound can have a very significant effect on sales. Marketing will want to be aware of the opportunities that improved SQ can provide in increasing product sales.

The job of Product Planning (PP) is to define those features that are desired in the product and to present those requirements to the Design functions. PP has to determine whether or not (with the help of Marketing) sound is important in the product, and if so, how the better sound is to be described to Design. A very common mistake made by PP in specifying a desired sound is to think that design needs quantitative specifications. The thinking goes "If a measurement of the sound pressure of our product is 74 dBA and the competition is measured to be 71 dBA, then our goal is to be at 70 dBA in the next model." The problem with this kind of "hard spec" is that PP has no way of knowing whether customers will think the new product at 70 dBA, if achieved, will sound better than either the old product or the competition's product.

A proper and more effective requirement from PP is a "soft spec" that specifies what PP really wants. A good example is one reported in the automotive industry as "Improve the sound of door closure without adding too much cost."[1] Although designers at first may protest that it is vague, it has the unquestioned value that it says exactly what PP wants. It can be up to the design functions through jury testing, signal processing, and mechanism analysis to determine what the "hard specs" need to be to achieve the goals set by PP.

Because this is not primarily a book about marketing and planning and most of the rest of this book is oriented to the activities in Design and Build, we will not dwell on those parts of the Benzene ring to any extent in this chapter. Product and Manufacturing Design (PMD) have experimental devices (frequency analyzers, modal analysis, acoustical intensity, etc.) and analytical procedures (FEA, BEA, dynamical modeling, SEA, etc.) for developing a product to meet PP's requirements. Manufacturing and Quality Assurance may have production line systems for product and/or process monitoring, control, and diagnostics for the Build function. We discuss such systems later in this book.

1.3
FOCUS GROUPS AND JURY TESTING

There is often some confusion about Focus Groups, used by Marketing to determine and rank product features, and Jury Testing, used by Design to set quantitative goals for product improvement. The role of these two activities in product definition will be quite different. In the case of a sewing machine for example, a Focus Group might be used to determine the stitches that need to be available, the slant of the needle bar (to make it easier to see the placement of the needle), or the type of foot-operated speed control. The Focus group might also be asked to evaluate and scale aesthetic features, such as color, the smoothness of surface finish, and form or shape. They can also be asked about sound, first to determine how important sound is as a product attribute, and also how they might describe sounds that are favorable or unfavorable. Such descriptions can be useful if a jury study is undertaken later.

A jury study is quite a different thing from a Focus Group. The Jury (or listening panel) Test should be thought of as a meter: stimulus in → response out. The purpose of the Jury Test is to determine the features of the sound itself that will be more favorable or less favorable to the acceptability of the product. A summary of Jury Test procedures is presented in Exhibit 1.1. The testing should be carried out under careful experimental design so that objective evaluation of the results can be found. The experimental design will include establishing the cohort (the population to be tested), defining the scaling method to be used, building the stimulus set, and establishing the trial protocol and the procedures for data analysis and presentation.

We shall delve into Jury Testing much more in a later chapter, but at this point we can make some general observations. First, it is very important that jury testing be conducted in a controlled manner. Because a perceptual response is being sought, there is a tendency to believe that the methods employed can be more casual, but in fact the opposite is the case. In the physical measurements that engineers are accustomed to, this discipline is built into the instrumentation and standard procedures. Jury testing, which is a part of experimental psychology, can be subverted by carelessness in the choice of subjects, the instructions given (or not given), and the test protocol.

1.4
THE ACOUSTICAL EXPERT IN THE DESIGN TEAM

Sound has gained attention fairly recently as a product attribute and as a result is "the new guy on the block" for product design. It has not been an easy matter for sound considerations to be integrated into the design process for both institutional and technical reasons. Certain turf battles between designers for lubrication, strength, and cooling have already taken place, and a *modus vivendi* of trade-offs between these needs has been established within the team. The acoustical engineer must somehow also become a trader in this process and therefore will upset the existing order to a degree.

Exhibit 1.1
The parts of a jury (listening panel) study of a product sound.

1. Selection of jury members
   - What is the population (the cohort)?
   - Are there special requirements (e.g., hearing)?

2. Training the jury members
   - in scaling methods
     (magnitude estimation, fixed interval, paired comparisons).
   - calibration using standard attributes (e.g., loudness)

3. Preparation of stimulus signals
   - component modifications
   - temporal/spectral modification
   - sequencing of stimuli

4. Design and conduct trials
   - form of presentation
   - (headphones, speakers, mono, stereo)
   - scheduling, briefing, monitoring of listeners
   - assembling results

5. Data analysis
   - scaling and normalizing
   - statistical interpretation
   - presentation of results

Compounding the problem is the fact that most design work related to product sound in the past has been Noise Control, which is after the fact add-on of devices to block or attenuate the sound. In the past, the product has been designed without concern for noise, so the acoustical engineer would not have been involved with the design team during the original design effort at all. Then, Noise Control was applied, rather like a bandage wrapped around the product to deal with the "unwanted noise."

As a result, the design team has been accustomed to ignoring the issue of sound because they expect the noise problem to be taken care of separately and after the fact.

The acoustical engineer must therefore find a place at the table in the basic design process. When design decisions are made that will affect the sound, they must be pointed out. If there is an alternative, that needs to be presented as an option. If there is another way to meet a need for structural rigidity, for example, without making the product louder, that needs to be presented. Product design is essentially an exercise in compromise, and the acoustical engineer must stand essential ground, offering alternatives whenever possible.

1.5
THE MEANING OF SOUND QUALITY

Because this book is about designing for sound quality, it is reasonable to expect the term to be defined. Such a definition has to be broad enough to encompass the many aspects of both sound and products, and yet specific enough to allow one to know when sound quality is in fact an issue. A suggested definition follows:

"Sound Quality (SQ) is the perceptual reaction to the sound of a product that reflects the listener's reaction to the acceptability of that sound for that product; the more acceptable, the greater the SQ"

The definition states that SQ is a percept, and it has therefore both subjective and objective parts. The percept relates to a particular product, therefore SQ is product specific. "A good refrigerator does not sound like a good lawnmower." And it also implies a scale for SQ in that listeners can make consistent judgments about SQ, which may then be incorporated into a physical scale for the component sounds for that product.

In this instance, the term *acceptability* is to be interpreted rather broadly. For products used in some situations, simple detectability of a particular sound may determine acceptability—the rattling sound of a loose fastener for example. In other products, speech interference may determine acceptability, whereas in others, appropriateness of the sound to the product determines acceptability. As we shall see, listeners may be asked questions about perceived power, how well made the product seems to be, how well it seems to be working, and its loudness. The general term of acceptability in the definition may refer to any or all of these, depending on the situation.

1.6
CHARACTERISTIC AND UNCHARACTERISTIC SOUNDS

The sounds that a product makes are an inevitable consequence of the components in that product. A fan makes some noise due to its blades moving in the air and its motor. As long as these sounds are in some balance with each other and not too loud overall, a user will be satisfied that it is working properly. The manufacturer may want to know how loud the product can be and still be acceptable, or what the range of motor and fan sounds should be relative to each other. As we shall discuss in the next chapter, those questions can be dealt with by the use of jury testing.

These sounds of motor and fan we would call characteristic of the product. They are expected, and unless too loud, the user is likely to accept them as characteristic of the product. But suppose the fan has a sound emanating from it that is unusual, such as a squeaking or clicking sound. The user will not regard such a sound in the same light as sounds characteristic of the fan. Such a sound might be the result of a problem that will lead to the failure of the fan or a danger of overheating. The user will have concern about using such a product.

Even if somehow the user becomes convinced that a breakdown or danger is not imminent, the sound will be attention grabbing and make the fan unsatisfactory as a product. It is not the *loudness* of uncharacteristic sounds that is the problem, it is their *presence* that is the problem. From an acoustical criterion point of view, it is the *detectability* of the uncharacteristic sound that is an issue.

Because of past work in psychoacoustics, it is possible to calculate how weak the offending sound must be for it not to be heard, so jury testing is not necessary in such a case. But of course, the manufacturer is likely to try to eliminate the sound by finding out why the uncharacteristic sound is present, not simply to reduce its loudness to inaudibility.

1.7
CAN THE SOUND ITSELF BE A PRODUCT DEFECT?

Most customers, manufacturers, and engineers would readily agree that a fan motor with bearings that squeak or grind because of a lack of lubrication would make uncharacteristic sounds and cause a shortened life of the fan. Such a product is defective, and the source of its problem can be traced to the defect in a part. The product is defective because it fails to meet the expectations of a reasonable customer with regard to function and aesthetics. The manufacturer will spend the effort to fix the problem so as to satisfy that reasonable customer.

But suppose there is some aspect of the bearings or motor shaft, not understood, that causes a squeaking sound. Extensive testing by the manufacturer indicates that the fan functions just fine as a fan; there is no indication of a shortened life or any danger in the use of the fan. Is the product still defective? Some engineers might say no, because the fan still functions. But the situation with the customer is exactly the same as it was when a bearing with a known defect was the cause of the sound. The customer will still be annoyed by the uncharacteristic sound, and concerned that the squeaking indicates early failure. The customer will be dissatisfied with the product because of its sound. From the manufacturer's point of view, the product is not satisfying the expectations of a reasonable customer. Clearly the product is still defective, simply because of the sound it is producing.

1.8
SUMMARY

For products to be designed and built with a sound that enhances their appeal to customers, all aspects of production may have to get involved: The desirable sound must be determined, the product must be designed to have the sound, and the product must be built so that the sound is actually realized in the finished product.

The acoustical specialist is a relatively new member of the product design team, and it will take some time to become a fully accepted member. Partly this is due to the history of "noise control," treated as an after-the-fact activity in product design. In part, it is also due to the lack of those handy "rules of thumb" that make it possible for specialists in other areas to make a quick evaluation of a design approach based on experience.

Manufacturers can access tools like jury testing and focus groups to set goals for product sound. In some cases, acoustical measurements alone may determine what the desired sound should be to achieve the desired sound quality.

Some parts of a product sound may be expected, and their loudness is the issue. Some sounds are uncharacteristic of the product. Uncharacteristic sounds are to be eliminated, i.e., made undetectable. Uncharacteristic sounds, if present, and characteristic sounds, if excessively loud, will cause a product to be rejected by reasonable customers. In such a case, the sound of the product is a product defect.

CHAPTER 2

Goal Setting Using Metrics and Jury Testing

This chapter is concerned with what the criteria for product sound should be and how they are to be evaluated. Historically, all such criteria have been based on personal interviews and structured listening tests. To the extent possible, the engineers and scientists concerned have attempted to correlate human judgments with physical measurements and algorithms. Such effects have been successful for many sounds, but there are many other judgments for which metrics are not reliable. In such cases, we have to retain jury testing as the ultimate tool.

2.1
WHAT ARE THE RIGHT GOALS FOR PRODUCT SOUND?

In Chapter 1 we discussed the major functional activities of a product company as presented in Figure 1.2. The functions were generally described as *Define, Design,* and *Build*. In Chapters 3 through 9 we will discuss in some detail how sound is generated and radiated by various mechanisms. Chapter 10 is of particular importance to the *Design* function as it is concerned with how measurements can be made on the production line to determine that the final products meet the goals of the designers and product planners, which of course is important to the *Build* function. It is the purpose of the present chapter to describe how we can arrive at the goals for the allowed sound from the various mechanisms in the product that is to be designed, and the SQ goals for the product sound as a whole is a part of the *Define* function. The purpose of this chapter is to set the context for the discussions that follow in the remainder of the book.

As indicated in Figure 1.2, the goals for the product sound are properly the province of the Marketing and Product Planning functions of the company. Marketing looks to the general population, its customers, and its dealers to determine how the company's products are accepted and what the competition is doing to meet customers' needs. Product Planning attempts to define the desirable features and functions of the future products to better meet customers' needs and desires. It is usually the job of Product Planning to tell the product design team what the next product should do and be through performance features, aesthetic characteristics, and price points.

Companies often use Focus Groups to assist Marketing and Product Planning to determine desirable features, function, and aesthetics. Focus Groups are also used to rank order different product attributes, e.g.; Is shape more important than color? Is sound more important than shape? Is user friendliness more or less important than product size? Properly structured, Focus Groups can tell a company how important sound is as a product attribute compared to other features and how customers relate to product sound—Do they use the sound in some way? Does the sound have only an aesthetic value?

There are certain aspects of product sound for which excellent correlations exist between various algorithms or metrics and the perceptual response to sound. These include loudness, detectability (or audibility), and speech interference. If the criterion or goal for product sound includes these attributes, then a calculation using standard metrics will usually suffice to determine whether a prototype meets the goal. We shall discuss these criteria in the following section.

We can think of loudness, detectability, and speech interference as criteria determined by the energy of the sound signal and how that sound energy is distributed in frequency. The advent of turbojet aircraft showed that a new dimension of the sound—annoyance—that was not determined by earlier criteria was important. A great deal of effort was expended to develop a metric for annoyance along the lines of the loudness algorithm. This procedure is described below. But unlike loudness, it was soon discovered that this algorithm was specific to turbojet aircraft and had to be modified to apply to turbofan aircraft, and it could not be used to predict annoyance from sonic boom or road traffic. We were able to see at that point that annoyance was (and is) specific to the product in question.

Sound also provides positive response to products, letting us know what is happening in the product or reinforcing our sense of quality. The car door closing sound and the sound of the motorcycle discussed in Chapter 1 are examples. Sound also identifies products so that we know if we are listening to a lawnmower or a dishwasher. Or, to turn the notion around, we expect that products of a certain type will have an appropriate sound. If a new product comes along, like an instant camera, that sounds very different from a quality 35-mm camera, there may be a problem with the acceptance of the product. These attributes of product sound have a cognitive aspect, in that the information carried by the sound affects how a customer will respond. None of the "standard metrics" will predict these attributes for products. In this area in particular jury testing is needed to determine sound quality.

2.2
ENERGY RELATED METRICS: LOUDNESS, DETECTABILITY, SPEECH INTERFERENCE, AND ANNOYANCE

There are many books on these standard metrics, so we will only describe each in terms of the methods used. There are two major methods of computing loudness, one due to Stevens[2] and the other to Zwicker[2]. We shall discuss the former. The basic difference between them is the way they deal with the phenomenon of masking, the reduction of sensitivity by noise in a stronger frequency band to noise in other bands.

The basic *quantity* of loudness is the *sone*, which is defined as the loudness of a 40 dB 1 kHz tone. The loudness of octave bands of noise is graphed in Figure 2.1, so that the loudness of each band of a sound spectrum can be determined (the loudness of a band of noise is greater than the loudness of a tone as indicated). These individual loudness values (called loudness indices), are combined in the Stevens method according to

$$S_{tot} = I_m + K\left(\sum I - I_m\right), \text{ sones} \tag{2.1}$$

where $K = 0.3$ for octave bands, $K = 0.15$ when the spectrum is presented in third-octave bands, and I_m is the index for the loudest band. The factor K represents the reduction in loudness of the other bands due to masking by the loudest band.

Loudness is also expressed in *loudness level*, expressed in *phons*. Loudness level is given by the sound pressure level of a tone at 1 kHz, that is equally loud to the tone being presented. The relation between loudness level and loudness is determined at 1 kHz and this is graphed in Figure 2.2. At sound levels greater than 40 phons, loudness doubles every 10 phons (doubles every 10 dB at 1 kHz) and doubles every three phons at lower levels. This relation is indicated both by the graph and the nomogram in Figure 2.2.

Figure 2.1   **Nomogram for calculating Loudness according to the Stevens procedure. This method is convenient for hand calculations, but is not so accurate in its treatment of masking as Zwicker's method.**

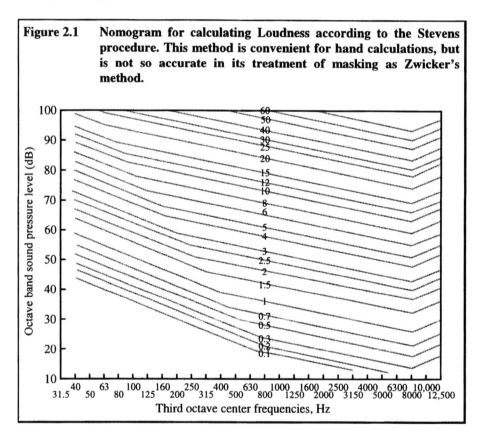

Figure 2.2 **Nomograms showing the increase of loudness as the sound level is increased and the relation between in sones and loudness level in phons.**

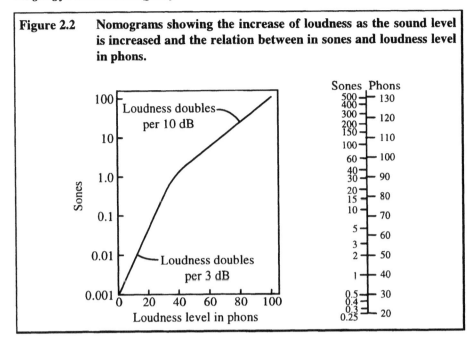

Speech Interference Level (SIL) is a metric indicating the ability of a noise background to interfere with the understanding of speech. The speech signal has frequency components from about 200 to 5000 Hz, and it varies in time with the natural rise and fall of the voice. Figure 2.3 illustrated how there is a variation of about 30 dB in the third-octave bands over that frequency range. Experiments at Bell Laboratories in the 40s showed that the contribution of these bands to speech comprehension was not equal[4]. The dots in the graph in Figure 2.4 each contribute an equal amount to speech intelligibility, so if the background noise spectrum covers 50 percent of the dots, then the fractional number left uncovered, the *articulation index*, is AI = 0.5. It is apparent that AI is a signal to noise ratio for speech in a noise background.

However, because there is a great deal of redundancy in speech, we can still understand over 95 percent of ordinary conversation even if the articulation index is as low as 0.4. For that reason, the allowed SIL for a background noise is set to the level at which it will produce an AI = 0.4. The levels for the 3 octave bands 500, 100, and 2000 Hz, which encompass most of the speech spectral energy, are averaged to give the SIL for the noise. Handbooks on noise control provide tables indicating acceptable values of SIL for various environments such as offices, restaurants, and aircraft.

Figure 2.3   **(a) Recording of third-octave band filtered speech shows how the dynamic range of speech varies with frequency.**

(b) Graphs of speech level show a 20–30 dB range, independent of the frequency.

(a)

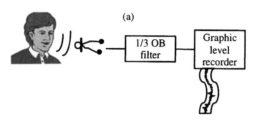

(b)

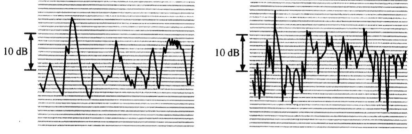

One-third octave band center frequency 125 Hz   One-third octave band center frequency 1250 Hz

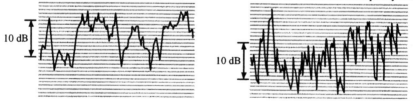

One-third octave band center frequency 250 Hz   One-third octave band center frequency 4000 Hz

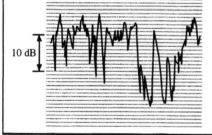

One-third octave band center frequency 630 Hz

Figure 2.4 The "dot counting" method of computing AI (articulation index) is popular with architectural acousticians working on office spaces.

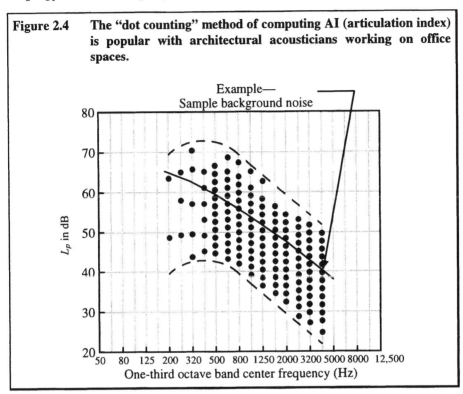

In addition to the ability of the auditory system to separate signals in frequency, it also separates signals in time, effectively integrating the sound energy over a time interval of about 25 msec to determine its loudness. If the third-octave energy spectra of a transient signal and the background noise are both integrated over such a period and if the energy of the signal exceeds the background noise energy, then the signal will be *detectable*. If the signal is expected (the next sound pulse from a dripping faucet), it will be detected at a level 5–10 dB even lower. This effect is termed "signal known exactly" and has its counterpart in purely electronic signal detection systems. More casually, we may speak of a listener becoming "tuned in" to a particular sound.

There is an additional metric that is energy based that combines the concepts of loudness and speech interference. This criterion is expressed as a group of spectra, known as *Noise Criteria* (NC) curves, shown in Figure 2.5. They were originally developed for application to quiet spaces such as offices and listening halls.[3] There are now several versions of these curves, differing primarily in their shape at low frequencies, but we can use the ones in this figure for our purposes. NC curves are of particular interest here because they have also been advocated as a way of evaluating the suitability of office products for noise.

The NC curves were proposed as a way to strike a balance between two objectionable characteristics of office background noise, usually due to air conditioning— *loudness* and *speech interference*. Some air conditioning background noise is desirable in office spaces to provide privacy between work spaces, but if it interferes with speech or it is simply too loud, it will be objectionable. Experiments involving interviews and jury studies of office workers have indicated that if the *loudness level* (LL in phons) exceeds the *speech interference level* (SIL in dB) by more than 22 units, then the loudness of the background noise will determine acceptability, but if $LL < SIL + 22$, then speech interference will dominate as a criterion. The shapes of the NC curves are approximately like an office background noise dominated by air conditioning noise, with a shape adjusted so that $LL = SIL + 22$.

Figure 2.5 **Noise Criterion (NC) curves combine criteria for loudness and speech interference. They were designed as criteria for the background noise in work and listening spaces.**

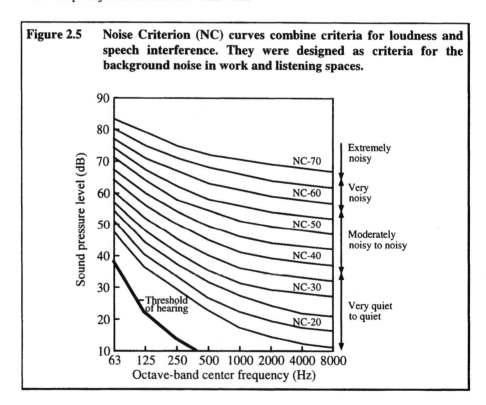

Some organizations have proposed that NC curves be used to rate the acceptability of office equipment, because they will contribute to the general background noise on the office. But this ignores the face that office equipment may produce sounds quite unlike the air conditioning noise for which the NC curves were developed. Office equipment produces tones, clicks, and other attention-grabbing sounds that are likely to make them more disturbing than the NC curves would indicate. Also, the sounds that such equipment makes relates to their functionality, and if those sounds are covered up by background noise, the usefulness of those sounds may be reduced.

When turbojet aircraft started to become widely used by the airline companies, it was noticed that the disturbance they caused did not correlate well with the predicted loudness, even though perceived and calculated loudness continued to correlate well. Respondents indicated they were annoyed by the sound of these aircraft, so a large effort was undertaken to develop a metric for annoyance. The Stevens method for the calculation of loudness became the model. The quantity of annoyance became the *noy* (the counterpart to the sone), and the counterpart to loudness level in phons became *perceived noise level* in PNdB.

The way that overall annoyance is calculated is exactly the same procedure as the Stevens procedure for loudness. The noise spectrum is plotted on a graph like that in Figure 2.6a. Each octave or third-octave band contributes a certain number of annoyance units to the overall value, but only the largest contribution is taken at full value. The remainder are reduced by the same factor as in Equation 2.1, and the overall annoyance rating in *noys* (the homonym is intentional) is calculated as before with the same conversion between noys and PNdB as between sones and phons.

Unfortunately, it soon became apparent that annoyance was not as robust a metric as loudness had turned out to be. When turbofan aircraft were introduced, they projected a highly tonal ~2 kHz sound forward on landing, which was very disturbing. An adjustment to the procedure for calculating annoyance was introduced that penalized the loudness level if a third-octave band in the spectrum exceeded the adjoining bands by a certain amount. The graph for this correction is shown in Figure 2.6b. This was the first indication that annoyance due to sound might be product specific.

Figure 2.6   **(a) Nomogram for converting a spectrum of jet aircraft noise to "Perceived Noisiness."**

(b) Correction to the Perceived Noisiness when a tone is present in the noise spectrum.

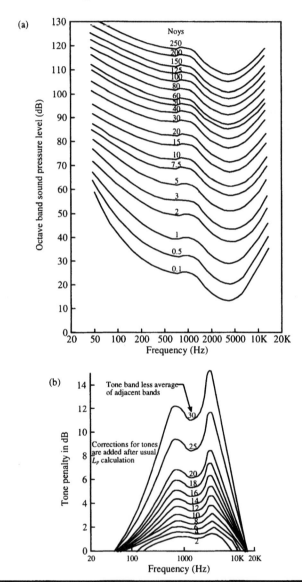

It became even more apparent that this annoyance metric was not universal when planners tried to apply it to predict disturbance due to automobile traffic and sonic boom. Perceived annoyance in PNdB simply didn't predict community annoyance due to these sources, and a complete alphabet soup of new metrics was introduced. All of this was just for the simple judgment of annoyance. Because customers' reactions to product sounds are related to more complex judgments than simple loudness or annoyance, other metrics or methods are seen to be needed.

2.3
THE CONNOTATIVE AND COGNITIVE ASPECTS OF PRODUCT SOUNDS —USING JURY (LISTENING) PANELS

We have noted that those aspects of product sound that are appropriate to the product ("a good refrigerator does not sound like a good lawnmower") and the sounds that give an impression of the manufactured quality or the operation of the product are not defined by the usual metrics that relate to loudness or annoyance. Although it is not definite that metrics will never assist in predicting such attributes of the sound, for the time being, having people listen to sounds and make judgments about the product must be a part of setting goals and evaluating products.

There are two main purposes of jury testing, One is product preference, i.e., which product is better made, would you purchase, is more powerful? The other purpose is to determine design features or degree of modification of a manufacturing process. The purpose of the testing will affect just about every aspect of experimental design and interpretation of the response data.

Jury Study Procedure

1) Exhibit 1.1 outlines the process of jury (listening) panel testing. The first step is to *select jury members*. There is some controversy on this point; in the automotive industry for example, "expert" juries are preferred, generally engineers who have detailed knowledge about and experience with the product. Such a jury is more likely to be able to associate a sound with a particular mechanism, but they can also be biased by their relation to the product and their knowledge of features in the product they may have been involved in. It is not likely that such juries reflect the attitudes and reactions of customers very well.

If customers are used for the jury, the first questions is, *Who are the customers?* They may be end users of the product, but they may be buyers for major chain stores, for OEMs, or for service outlets. Because such listeners do not have the experience of in-house engineers, the experimental design must be set up so that information about how the mechanism design can be improved can be gained. We shall see how this is done.

2) Because jury members are expected to respond to sounds in a structured way, they have to be *trained* in the meaning of the questions and in the scaling of responses. The attribute may be loudness (usually included for jury calibration), perceived manufacturing quality, or some other aspect of interest. These attributes have to be carefully explained to the jury so they know what they are expected to respond to.

   Popular scaling methods are *fixed interval*, (on a scale of 1 to 10, please judge...), *paired comparisons* (listen to the following two sounds and choose the one that ...), and *magnitude estimation*, to be described in the following point.

3) The *stimulus set* is the group of sounds the jury will hear. In the following section we will discuss how jury testing can help select among different potential designs by combining the sounds of the mechanisms that are in the product. This requires that the sounds of mechanisms be separated, which may be accomplished by both mechanical and electronic means. As an example, we will discuss how that is done for a vacuum cleaner.

   The duration of the sounds, the time between presentations, the sequencing of sounds, and other details of how sounds are presented are a part of the experimental design. Various designs, such as *central composite* and *factorial* are popular. We will discuss a test using the central composite design in the following.

4) The next step is to *conduct the listening tests*. The first issue is how to present sounds to the jury. Although headphones are convenient, there can be problems. Many consumers are not experienced in listening with headphones, and unless the sound is binaurally recorded, the auditory images will not be properly located in space. Loudspeakers provide a localized auditory image, which may be suitable, particularly for listening to appliance sounds, but not when the sounds surround the listener, as when one is seated inside an automobile. In such cases binaural listening is preferred. It is possible to provide binaural sound using loudspeakers for a single listener, but not for a group of listeners.

2.4
EXPERIMENTAL DESIGN AND EXECUTION USING MAGNITUDE ESTIMATION—VACUUM CLEANER EXAMPLE

The major sources of expected sounds from a vacuum cleaner are indicated in Figure 2.7. They are the (universal) electric motor, the main suction fan, the rotating brush, and the airflow through the machine. The experiment we shall use as an example will consider an array of machines in which these four sources are varied in strength. However, before we go into that example, note that we could also consider each of these sources in more detail. For example, there are several noise sources within the motor itself: the commutator/carbon brush interaction, the cooling fan within the motor, mass imbalance of the rotor, and the electromagnetic runout, which we shall discuss in Chapter 5. If our focus were on motor design as related to applications to vacuum cleaners, then these might be the sound levels we would change in the jury test.

The listeners will be asked to respond to the sounds with a number or letter. Highly trained listeners will have no trouble using a keypad or other electronic form of entry. Customers or users of consumer products may find it easier to write their responses on a sheet of paper. It is very important to monitor the subjects as this process goes on because confusions will almost certainly arise (is "good" a big number or a small one?) and it is important to catch such problems early in the experiment. Training and practice sessions help in this regard.

If the listener is asked to provide a preference between a pair of sounds (listen to the following sounds and select the sound that is more acceptable—a paired comparison test), then the response can be recorded on a paper sheet or electronically using a switch manipulated by the juror. Electronic response has several advantages because the interlocutor can spot erratic responses on the part of a listener, and the analysis of the data can take place without the step of converting paper responses to electronic form. An example of the screen display for such an automated system is shown in Exhibit 2.1.

5) The final step in the process is *data analysis and interpretation*. Fortunately, there is now a great deal of software available to assist in both experimental design and the analysis of response data. Some examples of the output of such programs are given in the examples that follow. In the examples, the analysis is used to construct a *response surface*, an objective and quantified multidimensional bilinear regression function that correlates the sound level produced by those mechanisms in the product that were varied in the stimulus set and are determined statistically reliable in the data analysis. This regression function allows one to anticipate the listeners' response (as far as sound is concerned) to a whole (virtual) world of products that might be designed, and from that group, to select one or more designs to be prototyped and tested for their sound quality.

Figure 2.7   **"Dirty Fan" and "Clean Fan" vacuum cleaners and their major noise-making components.**

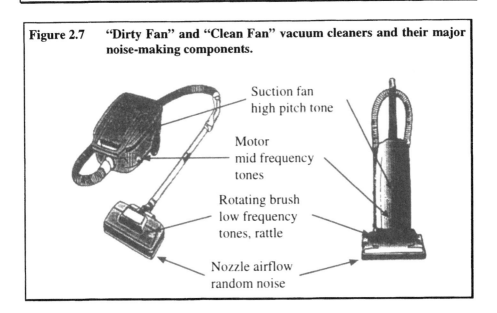

Suction fan
high pitch tone

Motor
mid frequency
tones

Rotating brush
low frequency
tones, rattle

Nozzle airflow
random noise

Exhibit 2.1

Computer screen display of the operation of an automated sound quality assessment system. The stimuli are presented to listeners and favored choices are recorded in a paired comparison procedure.

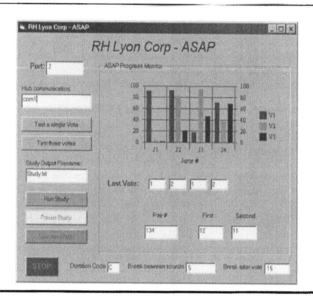

Experimental design places a practical limit on the number of variations we can allow in the stimulus set. The number of components to be varied will usually be no more than four or five, otherwise the size of the stimulus set (the number of different sounds to be presented) gets too large and the experiment cannot be carried out in a reasonable time.

The experimental design we describe here is a central composite design (CCD)[4], which is intended to anchor the judgments around a "standard," for example, the present product, and to produce a bilinear regression function for a judgment value as a function of step values in the sound components. We can illustrate the situation with the diagram in Figure 2.8. If we want to estimate a single variable functional form to the quadratic term, we need samples of the function at three locations at least, but if we want to both center the estimate on the 0 value and avoid the influence of the cubic term on our estimate, then we need to sample the function at steps $0, \pm1, \pm2$ as shown in Figure 2.8a.

Figure 2.8   **Sound level steps in a one- and two-component experimental design for the generation of a bilinear regression function for acceptability.**

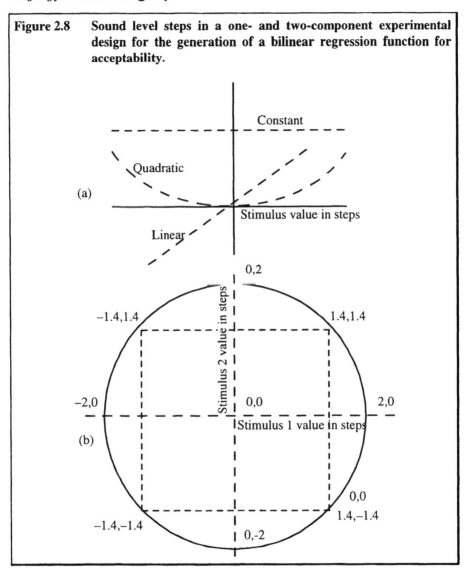

In the case of a two variable function, the sample points are as shown in Figure 2.8b and are chosen again to have samples at 0, ±1, ±2 from the origin, which means that the function is to be sampled at the points shown. The combinations of steps are therefore 0, 0; ±1, ±1; ±1, ±$\sqrt{2}$; ±$\sqrt{2}$, ±1; ±2, 0; 0, ±2, a total of 17 combinations, but not all combinations need be presented to the listeners. As the number of variables goes up, the number of combinations of steps increases. With four variables (our vacuum cleaner example), the number of combinations to be presented to listeners rises to 35. The sequence of stimuli is randomized and placed into blocks as shown in Exhibit 2.2. Each block contains the 0, 0, combination to lock in the judgments to the reference (e.g., the current product) stimulus, hence the term central in CCD.

In this case, the sounds of the components are separated primarily by special mechanical setups. For example, the sound of the rotating brush is determined using the setup shown in Figure 2.9. The brush is driven externally by a motor-pulley arrangement so that the cleaner motor is not operating and there is no airflow through the cleaner. The sound signal measured can be used directly as one component of the stimulus set, or the general spectral features of the brush sound can be used to produce a matched filter to electronically extract the brush sound signal from the overall vacuum cleaner sound.

It is also necessary to define the change in sound levels (in dB) that correspond to a step in the stimulus. If we expect that a ±10 dB range of variation for airflow sound is feasible, then each step for airflow sound could be 5 dB, whereas a step of only 3 dB might be chosen for motor sound because only a ±6 dB variation were deemed feasible for that component. Indeed, a step may not be quantitative at all, but simply refer to a type of component—a nylon versus a steel versus an aluminum gear, for example. In this case, there is no continuum in the component gear material, but the jury judgments can still be used to choose among them.

Exhibit 2.2
Sample set of sounds presented as a block to the jury. Each block contains the current or reference sound to keep the judgments anchored to the reference sound. This is part of a central composite design for the experiment.

SAMPLE/SOUND COMPONENT	MOTOR SOUND	SUCTION FAN	AIRFLOW NOISE	ROTATING BRUSH
A	−1	−1	1	1
B	−1	1	−1	1
C	−2	0	0	0
D	0	0	0	0
E	0	2	0	0
F	1	−1	−1	−1

Figure 2.9 **An example of "component modification": the sound level of the rotating brush (or agitator) of a vacuum cleaner is measured by driving it externally with a quiet drive.**

The type of scaling to be used, the number of listeners, and the duration of the test are not independent. Paired-comparison scaling usually requires a large number of listeners (100–200) listening for a few hours. Magnitude estimation and fixed-interval scaling require fewer listeners (20–40) listening for four to six hours. Because of the longer duration of these tests, they will likely be broken up into more than one session.

In the vacuum cleaner study, magnitude estimation was used for scaling, and the jury was trained by using geometric shapes projected on a screen. The jury was asked to scale the "bigness" of the shapes using whole numbers of their choice. At the end of the sequence, they were asked to write down their numbers for "extremely large," "very large," "not large," "small," and "tiny." By using these numbers, one can relate the scale of each respondent to that of the others. The same procedure was used later on the responses to the sounds of vacuum cleaners.

2.5
GATHERING AND INTERPRETING RESULTS OF A JURY TEST

If the product is a reasonably compact source of sound, like a vacuum cleaner, then loudspeakers are a natural choice for sound presentation. Posters, product samples, or other cues can be used to keep the listeners' attention focused on the product type as judgments are made (but care must be taken not to reveal brand names or other identifying information). The jury is instructed in the meaning of the attribute to be judged and how to score a response using magnitude estimation. Exhibit 2.3 gives instructions for *Effectiveness*, *Power*, and *Acceptability* for vacuum cleaner sounds. A response form for *Power* and *Acceptability* is shown in Exhibit 2.4.

Exhibit 2.3
Sample instructions to the jury for desired attributes to be judged. It is important that sounds be judged in the context for the nature and use of the product.

Instructions for Effectiveness: "... you will evaluate EFFECTIVENESS of the vacuum cleaners based on their sounds. To evaluate EFFECTIVENESS, listen to the sound and assign it a number that corresponds to how well it cleans, in your opinion. The better it cleans, the larger the number. The worse it cleans, the lower the number... Try to judge EFFECTIVENESS in the same way you would when listening to a blender, a dishwasher, or any other appliance, keeping in mind that each type of machine makes an appropriate sound."

Instructions for Power: "... you will evaluate the POWER of the vacuum cleaners based on their sounds. To evaluate the POWER, listen to the sound and assign it a number that corresponds to how powerful it is, in your opinion. The more powerful it is, the larger the number. The weaker it is, the lower the number... Try to judge POWER in the same way you would when listening to a blender, a car, or any other product, keeping in mind that each type of machine makes an appropriate sound."

Instructions for Acceptability: "... you will evaluate the ACCEPTABILITY of these sounds. Consider yourself to be a user of these vacuum cleaners in a home setting. You would have to listen to the sound of the machines while you clean for an extended period of time. Also, imagine yourself testing the vacuum cleaner in a store or listening to someone else cleaning in the same room while you did something else ... assign a number that reflects how acceptable the sound of the vacuum cleaner is ..."

Exhibit 2.4
Sample paper form used for judgments of vacuum cleaner power and acceptability, using magnitude estimation.

SPIN CYCLE

Name: _____   Date: _____

Time:_____ **RUN #1**		Loudness Scale—Run #1
		Please write down your numbers that correspond to the following judgments:
Block 1 **Loudness** **Gentleness**		
1. _____ 1. _____		
2. _____ 2. _____		☐ Extremely Loud
3. _____ 3. _____		☐ Very Loud
4. _____ 4. _____		☐ Moderately Loud
5. _____ 5. _____		☐ Quiet
Block 2 **Loudness** **Gentleness**		
1. _____ 1. _____		
2. _____ 2. _____		
3. _____ 3. _____		
4. _____ 4. _____		
5. _____ 5. _____		Gentleness Scale—Run #1
Block 3 **Loudness** **Gentleness**		Please write down your numbers that correspond to the following judgments:
1. _____ 1. _____		
2. _____ 2. _____		Extremely Gentle ☐
3. _____ 3. _____		Very Gentle ☐
4. _____ 4. _____		Moderately Gentle ☐
5. _____ 5. _____		Rough ☐
Block 4 **Loudness** **Gentleness**		
1. _____ 1. _____		
2. _____ 2. _____		You have completed RUN #1.
3. _____ 3. _____		Please turn to page 2
4. _____ 4. _____		and wait to begin RUN #2.
5. _____ 5. _____		
END OF RUN #1		

Once the data is collected from the jury responses, the correlation between responses and component sounds is determined through a regression surface, which attempts to fit the jury responses to a bilinear function of all the component sounds. There are a number of software programs that perform such an analysis. Exhibit 2.5 shows the output of one such program. The coefficients in the bilinear expansion are given, along with standard statistical parameters that indicate the reliability of the derived coefficient values and other information.

Exhibit 2.5

Typical printout of results of bilinear regression analysis of response data using a statistical package. Coefficients in the regression function are presented along with statistics showing the reliability of the estimate.

TUE   2/14/95   12:04:20 PM   C:\SYSTATW5\BEAT2.SYS

DEP VAR: ACCEPT   N: 30   MULTIPLE R: 0.972 SQUARED MULTIPLE R: 0.945
ADJUSTED SQUARED MULTIPLE R: 0.894   STANDARD ERROR OF ESTIMATE: 2.068

VARIABLE	COEFFICIENT	STD ERROR	STD COEF	TOLERANCE	T	P(2 TAIL)
CONSTANT	34.667	0.844	0.000		41.056	0.000
M	-0.979	0.106	-0.560	1.000	-9.277	0.000
S	-0.125	0.106	-0.072	1.000	-1.184	0.255
A	-1.021	0.106	-0.584	1.000	-9.672	0.000
B	-0.708	0.106	-0.405	1.000	-6.711	0.000
M*M	-0.076	0.025	-0.189	0.952	-3.060	0.008
S*S	0.018	0.025	0.046	0.952	0.739	0.472
A*A	-0.076	0.025	-0.189	0.952	-3.060	0.008
B*B	-0.052	0.025	-0.131	0.952	-2.110	0.052
*S	-0.016	0.032	-0.029	1.000	-0.483	0.636
M*A	0.055	0.032	0.102	1.000	1.692	0.111
M*B	0.055	0.032	0.102	1.000	1.692	0.111
S*A	-0.031	0.032	-0.058	1.000	-0.967	0.349
S*B	0.016	0.032	0.029	1.000	0.483	0.636
A*13	0.070	0.032	0.131	1.000	2.176	0.046

ANALYSIS OF VARIANCE

SOURCE	SUM-OF-SQUARES	DF	MEAN-SQUARE	F-RATIO	P
REGRESSION	1108.133	14	79.152	18.503	0.000
RESIDUAL	64.167	15	4.278		

From the designer's point of view, the important information is the degree of improvement available by changing the sound output of a component. Figure 2.10 shows how the *Acceptability* function, defined by Exhibit 2.5, depends on the airflow and motor noise components. In this case, it indicates that an equal reduction in both of about 5 dB is optimal. However, it might be much easier to change one component than another. If the motor is purchased and there is little leverage to get the supplier to change the design, then the design changes may have to look for other combinations of component reduction (in some cases an increase) to achieve the desired improvement. But the analysis will allow one to carry out such "what if" games.

Figure 2.10 **A response surface for vacuum cleaner sound Acceptability as it depends on airflow and motor sound levels shows a maximum when each component is reduced by ~5 dB. The scaling method used was "magnitude estimation."**

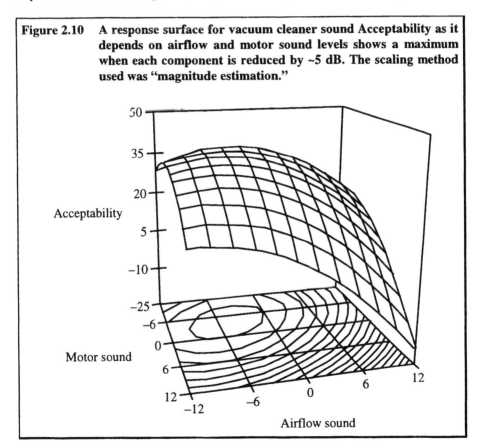

2.6
PREFERENCE TESTING—PAIRED COMPARISONS

Once one or more designs have been selected for potential exploration based on the jury responses, it may be desirable to test them against competing products, including the company's own current product. Because the sounds of the new designs are available, a jury test can be carried out to test an overall judgment, such as *Purchase Preference*, i.e., "Based only on the sound of the product, which one would you more likely purchase?" Such a test can use paired comparisons, with percentage selecting as the quantitative measure of preference. An example of the form used for such a test is shown in Exhibit 2.6, with the results for a washing machine test shown in Figure 2.11.

Exhibit 2.6

Paper form for a paired comparison test of preference of washing machine sounds.

WASHING MACHINE SOUND QUALITY ASSESSMENT
WASH CYCLE - I

Name _____   Session # _____

Check the box for the machine, A or B, that you would be more likely to purchase, based on its sound.

	Check				Check	
Pair #	A	B		Pair #	A	B
1.				12.		
2.				13.		
3.				14.		
4.				15.		
5.				16.		
6.				17.		
7.				18.		
8.				19.		
9.				20.		
10.				21.		
11.						

Figure 2.11   A comparison study of two new designs of a washing machine with other machines on the market for the wash and spin parts of the cycle used the paired comparison scaling procedure.

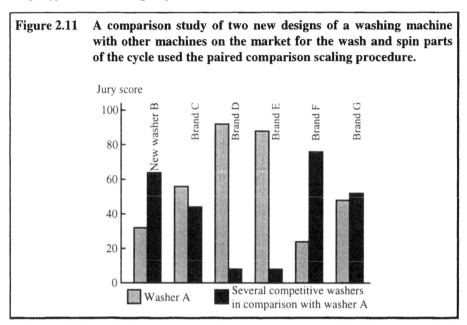

In the washing machine preference test, there are two potential designs tested against each other and against other products on the market, including the company's own current product. This comparison can give the company very useful information on how to present the product with regard to sound. In some cases, the producer may only care about the comparison of two products. Even in this case, it is preferable to place the two product sounds of interest into a group of ten or so competing product sounds so that larger scale statistics are achieved and the likelihood of listener boredom is reduced.

2.7
IMPLICATIONS FOR DESIGN

The preferred design, according to the process just described, is a new combination of the loudness of component sounds within the product. For that reason, we concentrate on the relation between component design and component sounds in the following chapters. There is some controversy regarding this approach. Some perceptual psychologists, among which psychoacousticians are a subset, propose additional metrics that can be used to choose among product variations. In the area of sound, such metrics carry names like roughness, sharpness, and fluctuation strength. They are measured using combinations of frequency and temporal filtering, and instrumentation is available for computing these metrics.

These metrics have undoubtedly shed some light on the correlation between features of the sound and perception. But engineers design gear trains, motors, and structures, not spectra, so a correlation between component sounds and the acceptability of a product (which we have defined as Sound Quality) is of more direct value to the design engineer.

CHAPTER **3**

Noise Control, Noise Reduction, and Design for Sound Quality

In this chapter we discuss the mechanical analysis part of our approach to designing for product sound quality in more detail. Once the criteria for the product sound have been established as discussed in Chapter 2, we need to proceed to the actual physical design of the product that will meet the goals. What are the things the acoustical specialist needs to think about as the product design evolves, and how can his/her expertise solve sound quality problems consistent with other nonacoustical requirements?

3.1
THE TRADITIONAL ROLE OF NOISE CONTROL IN PRODUCT DESIGN

For about the past half century, the role of acoustics in product design has been directed at reducing the sound radiated by noisy products. These products, primarily used in the manufacturing, transportation, and construction industries, have an impact on the general population, users and nonusers alike, and their sound has become a regulatory issue. This history has had a great effect on how product sound is viewed by the acoustics profession, manufacturers, and the public.

For regulatory purposes, the metric of sound to be used should be as simple as possible to measure and to design. For both purposes, the A-weighted sound pressure level $L_p(A)$ turns out to be quite a good choice because it is easily measured with a sound level meter, and it correlates well with loudness, speech interference level, and potential for hearing damage for a variety of products. The fact that $L_p(A)$ does not reveal anything about the information content of the sound being measured is not a disadvantage for regulatory purposes because an enforcement official may not be that concerned about the "intent" or sound quality of the noise producer.[5]

From a design viewpoint, concern for $L_p(A)$ as a metric places an emphasis on reducing the loudness of the product without concern for other aspects of the product sound that were discussed in the preceding chapter. As regulatory issues often arise *after* the product is in the market, the use of add-on devices for noise reduction has been a natural consequence of the passage and enforcement of such laws.

Exhibit 3.1 contrasts the differences between the traditional approach of noise control with designing for sound quality. Noise control methods are well documented, and many textbooks and handbooks provide good information regarding the design of noise enclosures, isolation systems, silencers, and the other components of a noise control design. Because these are add-on devices, they are fairly independent of the basic product design, and it is therefore fairly easy to say what the noise reduction effort has cost. In fact, one often hears the phrase "dollars per decibel" to indicate that such a cost tradeoff is both possible and natural. But the costs of add-on devices is a variable cost of manufacturing—it is paid for in every product that is shipped.

Most of these features of noise control may be considered advantageous, but the enclosures, barriers, and other elements of the process may reduce the reliability or maintainability of the product. It is not unusual to visit a plant where noise enclosures have been installed to see that the maintenance staff has removed some enclosure panels for access to the machinery. Reaching into machinery areas where vision is restricted by a noise control panel can be dangerous.

Designing for sound quality, by contrast, depends on understanding some of the basic principles encountered in this book. Unfortunately, many persons who have the understanding to deal with sound at the design stage are not on the product design teams, but are in the corporate research laboratory.

Exhibit 3.1

A comparison of the activities and consequences of efforts in the traditional noise control of products, contrasted with activities in design for product sound quality.

COMPARISON: NOISE CONTROL
AND DESIGN FOR SOUND QUALITY

Noise control	Design for sound quality
■ Technology is in place	■ Principles are understood
■ Implemented by handbook	■ Practitioners are scarce
■ Add-on after product design	■ Integrated with design cycle
■ Independent design	■ Highly interactive
■ Costing is straightforward	■ No separate costing
■ May reduce reliability, maintainability, safety utility	■ May enhance function or other attributes

As a component of the basic design effort, the acoustical design components must be integrated at an early stage and are highly interactive with other aspects of the design such as cooling, lubrication, flammability, and electrical integrity. Because of this highly integrated nature of design it is not possible to say exactly what it costs to design for a "good sound," any more than it is possible to separate out the costs for the lubrication, structural, or cooling parts of the design. In this sense, an indication of "dollars per decibel" costs may signal that the design was done after the fact and not properly integrated into the basic design.

Because acoustical considerations interact with other design components, it is always possible that extra benefits may come from incorporating acoustics in the design. A machine cover that is perforated to reduce its radiation efficiency may also enhance cooling, leading to other benefits. Or increases in cooling may allow a larger motor and greater torque, leading to reduced gear noise. Because design components interact, including acoustics in the design can in some cases reduce costs or enhance other attributes.

3.2

THE GRAND STRATEGY: EXCITATION, RESPONSE, AND RADIATION

Our approach to acoustical design is to understand the process of sound production in products through a three-part sequence, a triptych. This concept is illustrated by the topics in Exhibit 3.2. First there is excitation, the generation of high-frequency forces and motions that are the sources of high-frequency energy (we'll call it the noise energy), some of which ends up as sound. These arise from the fundamental mechanisms in the product: electromagnetics, combustion, and gearing. We shall consider a number of such sources of excitation that commonly occur in various products.

Exhibit 3.2
The design triptych shows options for redesign of sound and blocking of noise energy in different parts of the product.

ELEMENTS CONSIDERED IN DESIGN

Generators ("it's better to control at the source")

- AF energy is small fraction of machine work
- Little vibratory energy is radiated
- Modifying source may interact with function

The transmission path

- Made up of assembled parts
- Offers opportunity for design changes
- Available methods:
    - impedance mismatch
    - isolators
    - loading masses (spermassen)
    - damping

Radiating surfaces

- Reduce vibration by isolation
- Dissipate vibration by damping
- Decouple from air by composite layers
- Decouple from air by perforation

Because the noise energy is concentrated at the source, if we can find a way to modify it there, that will be an effective way to reduce or modify the sound. Once the noise energy spreads out into the structure and the air, it is more difficult to control. We might select gearing that is more accurate or modify a cam profile to accomplish that goal. But such a strategy has problems if the modification affects product performance, affects efficiency, or has an adverse environmental effect. Changing the fuel injection profile of a diesel engine can reduce combustion-induced noise, but it may also reduce horsepower and increase chemical pollutants.

Secondly, the forces and motions of the sources produce vibrations, both directly where they operate and more remotely as a result of the transmission of noise energy within the product—in air passages and through structural connections. Because the amount and the frequency content of this energy are modified by passage through the product, this transmission path can be an important place to look for benefits from redesign. Very often the path of noise energy transmission is made up of assembled parts (the connecting rod and crankshaft of an engine for example), and such elements are more readily modified as one progresses from one design or model to the next.

Noise energy in the transmitting components (structural or airborne) can be modified by dissipation (vibrational damping or sound absorption), or by reflecting the energy to keep it from reaching openings or structures from which it can radiate. If the transmitting structure is fairly light and flexible, added damping treatments can be effective if properly designed. Stiffening of panel structures makes conventional damping treatments less effective because a smaller fraction of the potential energy of vibration is stored in the damping treatment, and the treatment only dissipates a fraction of the energy that it stores. Some newer damping systems specifically designed to damp high-impedance structures are also a possibility.

High-mobility junctions (compliances) are effective when the joined structures are of low mobility (heavy and/or stiff). This is the ordinary situation in which one uses a flexible isolator between a machine and its foundation to reduce the force transmitted to the foundation, low-mobility junctions (loading masses or "spermassen") are effective when the joined structures have high mobility (light and flexible). One of the most difficult situations is where a heavy (low mobility) frame supports a lightweight flexible (high-mobility) machine cover. An isolator that is soft enough to reduce vibration transmission will likely be too soft to properly support the cover. If the edge of the cover is framed, then the mobility of the cover may be reduced enough so that an isolator between the machine frame and the cover edge can be effective.

When some of the noise energy reaches the external surfaces of the product where it can be transmitted into the air as product sound, we have other means of modifying the sound. Sound radiation is in effect a product of the vibrational amplitude of the structure (e.g., measured in ms velocity) and the coupling between the structure and the surrounding air (measured by the radiation efficiency). We can modify the sound by dealing with either or both of these factors. The vibration can be reduced by keeping it out of the housing through decoupling (dealt with above) or by dissipation of structural vibration (damping).

The radiation efficiency is reduced by adding a commercially available decoupling composite to the structure, or by "opening up" the structure (both are discussed in Chapter 9). This may be done in some circumstances by perforating the structure or by replacing a solid sheet with expanded metal or a truss structure. Such a structure may vibrate, but air will pass through it rather than being compressed and producing a sound wave. If such openings in the panel are allowable, then this solution can be very attractive, as it is effective and reduces material usage and product weight.

3.3
SOURCES OF FORCE AND MOTION

In electrical systems we are accustomed to thinking of sources where the voltage drop produced is (nearly) independent of the load applied (battery, wall outlet). We speak of such sources as "prescribed drop" sources having low internal impedance. Other, less familiar, sources produce a current flow or charge displacement that is nearly independent of load (piezoelectric transducers for example). We speak of these as sources of prescribed flow having high internal impedance.

In mechanics, there are sources of vibration that tend to produce a certain motion (cams, gears) regardless of the load, whereas others tend to produce a certain force (electromagnetics, airflow) regardless of load. In a mobility diagram describing a mechanical system, motion (velocity) is the "drop" variable and force is the "flow" variable. We therefore speak of sources like a cam interface as a source of prescribed drop and turbulent fluid pressures as a source of prescribed flow.

This concept is important because it gives us guidance in how to modify the output of these sources. If we want to modify the force output of a source of prescribed motion (low internal mobility), we will connect it to a load of high mobility, just as we would reduce the current drawn from a battery by connecting it to a high resistance. If we want to reduce the force output from a source of low high internal mobility (electromagnetics), we will connect it to a load of low mobility (a massive or very stiff structure).

The discussion of sources of noise energy in the next two chapters follows this logic by first discussing sources of prescribed motion and second sources of prescribed force. But, just as sources are never pure in electrical systems (batteries always have some internal resistance), the breakdown into prescribed motion and force is an idealization, assisting us in thinking about induced vibration and sound, but not an absolute prescription in any particular situation.

3.4
STRUCTURAL RESPONSE TO THE SOURCES OF NOISE ENERGY

Chapters 6 and 7 describe the structural response to the mechanisms that are sources of force and motion. Chapter 6 deals with the single degree of freedom (dof) resonator that is a good model for a machine supported on its foundation by isolators. It turns out that much of the design of isolators can be determined from such a simple model. But more importantly, the resonator is a useful model for any of the many structural resonances that occur in a product in the audible frequency range. The parameters of the single dof resonator are its mass, stiffness, and damping. These same parameters are also important for much more complex structures.

Chapter 7 describes the vibrational response of structures in terms of the kinds of waves and resonances that are excited by the sources. Different wave types (flexural, longitudinal, shear) participate to varying degrees in each of the resonances of the structure. It is the presence of these resonances that allow the structure to absorb noise energy from the sources and to transmit that energy to other parts of the product. It is remarkable that no matter how complicated the structure is, each of these resonances can be represented by the single dof resonator that we discuss in Chapter 6.

3.5
SOUND RADIATION FROM THE PRODUCT

As we shall see, the forces and motions of the noise energy sources in some cases radiate sound directly into the surrounding air. But at least as often, it is the structural vibration discussed in Chapters 6 and 7 that radiates the sound. In these cases we need to predict or measure the ability of the structure to radiate sound. That topic is covered in Chapters 8 and 9.

Certain motions of the sources and the structure can be considered to be "elementary" sound sources. These are discussed in Chapter 8. They include the pulsation (volume pumping) of the source and sudden accelerations and decelerations of a component. Because these kinds of sound radiation are very sensitive to the rate of pulsation and acceleration, emphasis in reducing the sound usually entails slowing down the motions and forces as much as possible.

Chapter 9 considers sound radiation from more complex situations. Extended structures will have a number of parameters that control the amount of radiated sound. For vibration, these are combined into a "radiation efficiency," and we are usually trying to modify this parameter in design. When fluid forces radiate sound, other parameters are significant. Fundamentally they are related to "dynamic head" and the "Strouhal frequency" of the flow field, but practically they can be expressed in terms that relate directly to the product, such as fan tip speed, volume flow rate, and dimensions of blades and air passages.

The measurement of product sound is an important part of the design activity because at critical points in time we need confirmation of our sound predictions as we proceed with the design. Three methods of sound measurement are described in Chapter 9: reverberation room, sound intensity, and hemi-anechoic room methods.

3.6
QUALITY ASSURANCE (QA)

The book concludes with Chapter 10 on QA methods that use sound, vibration, and other signals to determine that the product meets SQ goals as early in the production process as possible. It is difficult to generalize concerning the methods to be employed to accomplish this because of the very specific situation in different production facilities, even for the same type of product manufactured by different companies. For that reason, we can only provide some examples in this chapter to indicate the general approach and some specific case studies.

CHAPTER 4

Sources of Prescribed Motion: Imbalance, Cams, Gears, Chains, and Sprockets

As indicated in the last chapter, certain mechanisms used in products tend to produce a motion that is nearly independent of the load presented to them. This is a natural part of the design of gears and cams because they are shaped to achieve a particular type of motion in their follower. Although not totally independent of load, their prescription as "sources of prescribed motion" is a useful classification.

4.1
RELATION BETWEEN DESCRIPTIONS OF THE SAME SOURCE

In the last chapter, we discussed why the internal mobility, compared to the load mobility, determines the actual output of a mechanical source. We can illustrate that concept and an equivalence between sources of motion and sources of force by considering the imbalance of a machine. Figure 4.1 shows a machine sitting on some springs and vibrating with displacement y as a result of the motions of its internal parts. The figure indicates we can consider this situation equivalent to a superposition of two situations: one in which the internal mechanisms operate and an external blocking force L_{bl} is applied to stop the vibrations, and one in which the internal mechanisms do not operate and $-L_{bl}$ is applied to the machine to produce the displacement y.

Figure 4.1   **The action of internal moving parts of a machine can be represented as equivalent to an external that is required to block or keep the machine from moving, or equivalently, an external force required to produce the same vibrations when internal components are quiet.**

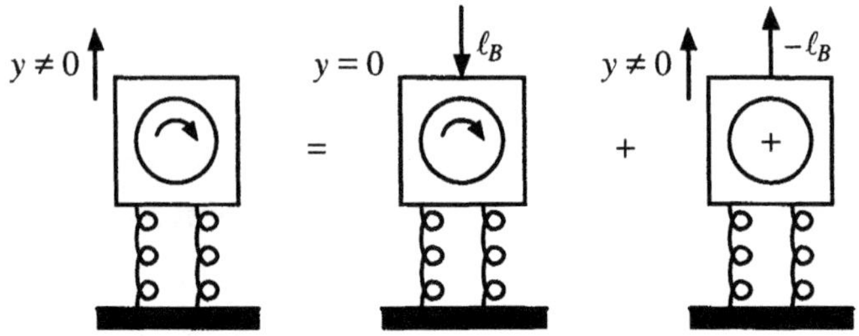

If we assume that the operating frequency of the machine (of mass M) is higher than the mass-spring resonance frequency of the machine sitting on its springs (the usual case if the spring is to act as an isolator) then the vibrational velocity of the machine V is related to the force by $L_{bl} = j\omega M V$, which corresponds to an internal mobility $Y_{int} = 1/j\omega M$. This can be represented by either of the diagrams in Figure 4.2. Figure 4.2a represents the imbalance as source of prescribed force L_{bl} in series with the internal mobility Y_{int}, and is the *Thevenin* representation of the source. The diagram in Figure 4.2b represents the imbalance as a source of prescribed velocity $V_{free} = L_{bl}Y_{int}$ where V_{free} is the vibrational velocity the machine would have while operating freely suspended. The diagram in Figure 4.2b is referred to as the *Norton* representation of the source.

The diagrams in Figure 4.2 show that the free velocity V_{free} will be imposed on a load when the load mobility Y_{load} is large compared to Y_{int}, whereas the force L_{bl} will be imposed on the load whenever Y_{load} is small compared to Y_{int}. Therefore we see that the source of imbalance may be thought of as either a source of prescribed motion or force, depending on the relative values of internal and load mobilities. Nevertheless, we classify this source as one of prescribed motion because as a famous professor from MIT has said, "the center of gravity (CG) wants to stay in the same place".[6] If the internal machine components are moving in such a way as to change the location of *their* CG, then the outer housing must move in such a way as to change *its* CG. Housing displacement is therefore the fundamental variable that describes this source.

Figure 4.2   A mechanical source of vibration may be represented as
(a) a Thevenin source (prescribed drop), or
(b) a Norton source (prescribed flow).

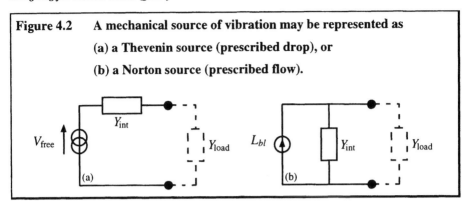

4.2
EXAMPLE OF RECIPROCATING IMBALANCE—A SEWING MACHINE

The procedure just described allows us to represent the imbalance either by free motion or by blocked force. It also allows us to measure one and compute the other. Figure 4.3 shows a sewing machine suspended by soft elastic cords to make sure its motions are mass controlled. Imbalance forces are at the operating speed (10–15 Hz) and a few multiples (harmonics), while the lowest structural resonance of the machine is more than 100 Hz. At the operating frequency, the machine is dynamically just a lump of mass with a CG, total mass, and moments of inertia about its principal axes.

This particular machine has a needle bar that is inclined. Vertical acceleration, converted to blocked force by machine mass, is shown in Figure 4.4, and fore-aft acceleration converted to force is shown in Figure 4.5. The vertical acceleration shows a large fundamental component of force with a smaller second harmonic. The fore-aft component has a larger second harmonic, but with some fundamental.

We can convert the horizontal and vertical force data in Figures 4.4 and 4.5 to a *polar* description as in Figure 4.6. This diagram shows more clearly the angled reciprocating imbalance of the inclined needle bar and some small "ears" of imbalance force at one end of the cycle. These ears are due to imbalance in the take-up, the four-bar linkage that pulls the thread tight after each stitch is made.

When the wheel of a car is out of balance, we solve the problem by balancing the wheel, adding a bit of mass to the rim. Can we balance the needle bar of the sewing machine in that way? Figure 4.7 shows that a reciprocating mass, oscillating in a horizontal sinusoidal motion, can be resolved into two counter-rotating masses of half value. If the shaft driving this mechanism rotates in a clockwise direction, then one of the rotating half-masses can be balanced, leaving a half-mass rotating in the counterclockwise direction, which cannot be balanced by a mass on the drive shaft. If a full mass is used for balancing in the clockwise direction, then a reciprocating mass in the *vertical* direction results, which might be advantageous, as discussed later in the book.

If we want a full balance for the reciprocating imbalance, then a *counterrotating* mass must be introduced, which requires a counterrotating shaft in the machine. Such shafts have been included in some engines. Single-cylinder engines will have a counterrotating shaft at the same rate of the main shaft. In four-cylinder engines, an imbalance at *twice* the main shaft rate is residual, and a shaft counterrotating at twice the main shaft rate is introduced to supply the balance mass. The Chrysler *Silent Shaft* engine was an example of such a design.

Figure 4.3   Free motion due to imbalance can be measured by suspending the product on soft elastic cords during operation.

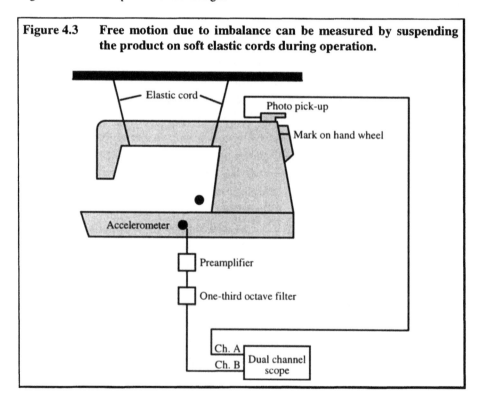

Figure 4.4 The vertical imbalance force is inferred from the measured vertical acceleration and the product mass. A small second harmonic is evident.

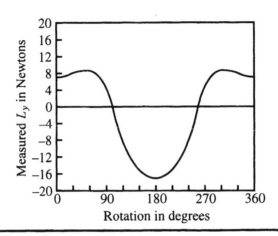

Figure 4.5 Fore-aft imbalance force determined from the acceleration shows a large second harmonic with small fundamental.

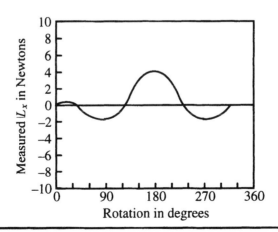

Figure 4.6   **Combining vertical and fore-aft forces into a polar diagram shows the slanted imbalance due to the slanted needle bar. The small "ears" on the diagram are due to the take-up.**

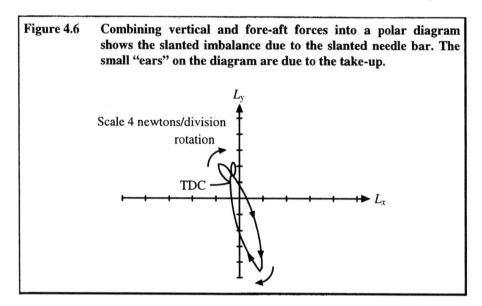

Figure 4.7   **A reciprocating imbalance at the shaft rotation rate can be resolved into two counterrotating imbalances of one-half the value.**

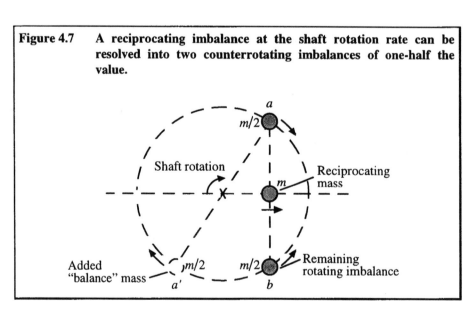

4.3
GEAR MESHING AS A SOURCE OF PRESCRIBED MOTION

If we wanted to transmit rotation from one rotating shaft to another, we might think of unwrapping a belt or rope from a wheel on one to a wheel on the other, as shown in Figure 4.8. Such an arrangement should transmit power smoothly if the belt does not stretch and the wheels are truly round. If we follow a point on the belt as it unwinds from one wheel or winds on to the other, a curve is described that is called the *involute* of the circle.

Figure 4.8 **A point on a belt that joins two pulleys follows an involute path as the belt unwinds from one wheel and winds onto the other. This is the basis for the involute gear shape.**

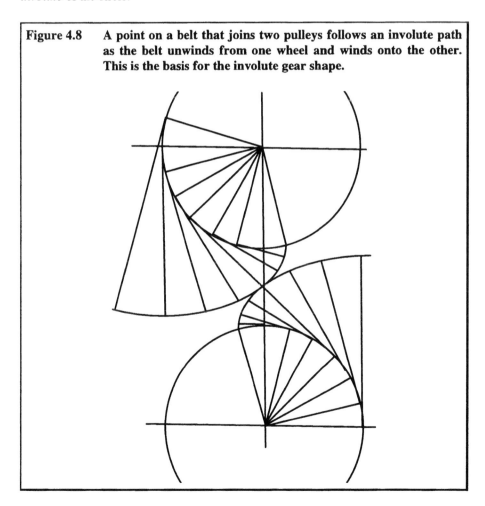

If we create two mechanical surfaces in contact that have the involute shape, and if these surfaces do not deform under the load, then the point of contact between them is like the point on the belt, and power is transmitted smoothly between the two wheels as before. These mechanical surfaces become the faces of an *involute gear*. If the driving wheel turns at a constant rate of rotation, the driven wheel will also turn at constant speed. If the mating teeth are not true involutes, then a constant rate of rotation in the driving wheel will result in fluctuations of speed in the driven wheel. These fluctuations are referred to as *transmission error*.

The sources of transmission error, therefore, are essentially geometric and include errors in the fabrication of the gears and the effects of deformation. As illustrated in Figure 4.9, fabrication errors will include profile and pitch variations in straight cut spur gears and variations in the theoretical helical line of contact (lead) for helical gears. As a result of the long history and importance of gearing systems, there is an enormous technological base of information and practice of gear design that we cannot go into here. However, the idea of transmission error as a motion source is valuable to us, whatever the cause of such errors may be.

Figure 4.9   **Meshing gears can have transmission error for a variety of reasons including noninvolute geometry, tooth bending, and pitch variations.**

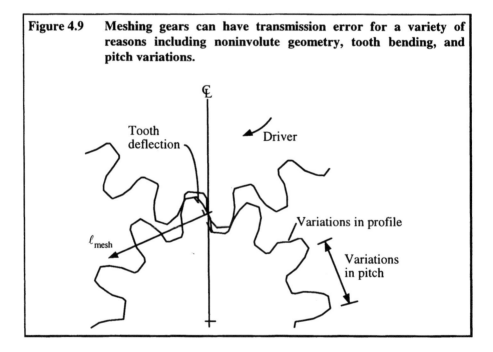

Transmission error as a source of vibration can be represented as shown in Figure 4.10. This figure assumes that the transmission error results only in fluctuations in torque and angular velocity, although, of course, lead errors will also result in fluctuations in force along the axis of rotation. But the simple one-dimensional diagram of Figure 4.10 will provide us with the needed concepts.

Although the transmission error is mutual we can think of it as a source of motion V_{te} of gear one with internal mobility Y_1 driving the "load" mobility Y_2 of gear two (or vice versa). In Figure 4.10, we regard transmission error as a source of motion (velocity), which makes sense from its basic cause of geometric variations. However, if the loading gear two had very small mobility, we could use the Thevenin-Norton equivalence of Figure 4.2 to convert transmission error to a source of prescribed force $F_{te} = V_{te}/Y_1$.

The errors in gear teeth depend on the manufacturing process. A manufacturer of industrial forklift trucks purchased experimental gears resulting in the noise spectrum shown in Figure 4.11a. The peak at 1 kHz is the meshing frequency of a gear pair that was causing a noise problem, and the noise spectrum in the figure was considered satisfactory. A larger number of gears of the same specification for accuracy was ordered, but to the manufacturer's chagrin, the noise from these gears was unacceptable. As we can see from the noise spectrum for the second set of gears, shown in Figure 4.11b, the difference between the two sets is a strong peak in the acoustical energy from these gears at about 3 kHz, the third harmonic of the mesh frequency.

Figure 4.10   **Transmission error is thought of as a source of prescribed motion. The mesh force is determined from the mobilities of the gear teeth at the interface.**

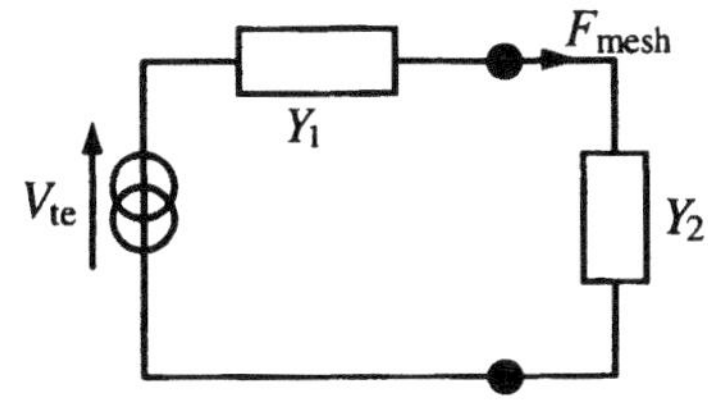

Figure 4.11   **(a) A frequency spectrum of the radiated sound of quiet gears in a forklift shows a mild peak at the gear mesh frequency at 1 kHz for ground gears.**

(b) Hobbed gears of the same class rating show a strong and objectionable tone at 3 × gear mesh frequency.

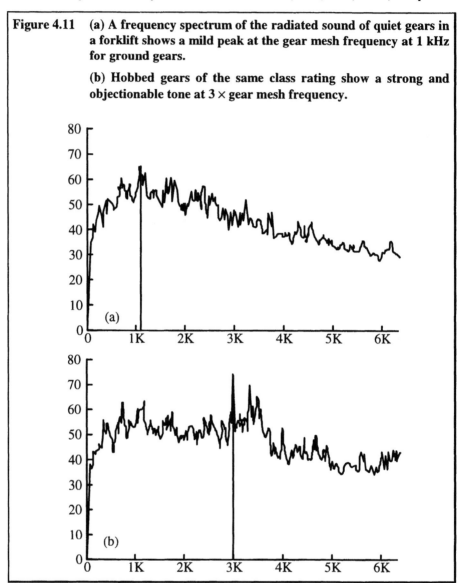

Naturally, the manufacturer thought that the new gears were not as accurately cut as they should be. Gear accuracy is determined by an instrument that measures the profiles of individual teeth. The output from this instrument for the (noisy) gear teeth in question is shown in Figure 4.12a for the forward and rear profiles of several teeth. A perfect involute profile would be a horizontal line midway between the dashed lines. The actual profiles lie between the dashed lines so the gears do in fact meet the required specification for accuracy. But if we examine the profiles more carefully, we see that the variations in profile for all the teeth are very similar, an observation reinforced by overlaying all of the profiles as in Figure 4.12b. The variations also show an antisymmetric pattern in their shape, which is probably the reason for the strong third harmonic in the mesh tone.

The reason these gears, which had the same accuracy rating as the experimental gears, were noisy is most likely this repeated pattern of the transmission error from one gear to the next. The pattern did not occur on the experimental set because for a small production run the gear manufacturer used a grinding process to make the gears. Such a process tends to introduce random errors in the profile. For the production run, the gears were cut by hobbing, a machining process that tends to repeat errors in the profile.

4.4
SPROCKETS AND CHAINS—AN EXPERIMENTAL DRIVE SYSTEM

Figure 4.10 indicates that to reduce the force that flows in the diagram due to the transmission error, we need to increase the junction mobility $Y_1 + Y_2$ of the mesh. Physically, we can understand that to allow the geometric variations to be accommodated without creating large forces, the junction has to give a bit. If I drive my car over a rough road where the tires have to remain in contact with the road surface, I may let some air out of the tires to increase their compliance and to reduce the forces that cause the car to jostle me. The increase of compliance (by reduced tire pressure) produces an increase in the mobility at the tire-road interface.

Figure 4.13 is a sketch of a drive system for a prototype copying and reproduction product consisting of a motor, timing belts, gears and a chain drive with several drive sprockets, and idlers that serve to redirect the chain. A spectrum of the noise from this setup is shown in Figure 4.14. It consists of a group of tones below 1 kHz and a region of broad band noise due to the chain drive above 1.5 kHz. It is the noise from the chain and sprocket components that we pay attention to here.

Figure 4.12   **(a) A gear face profile recorder confirms that the hobbed gears are in the specified class.**

(b) An overlay of the gear face profiles indicates that the hobbing process put nearly the same error into every gear tooth. Grinding of gears tends to put more random errors into the teeth. The antisymmetric nature of the profile error is probably the reason for the 3× mesh.

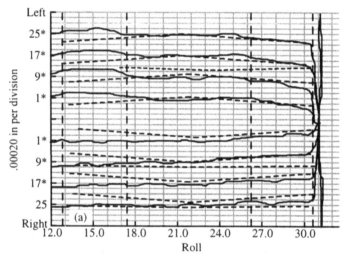

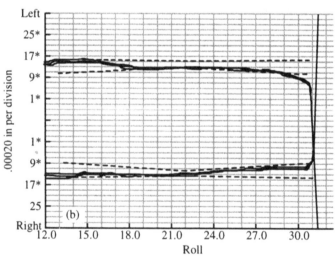

Figure 4.13   A structural casting holds motors, gears, sprockets, chains, and timing belts in alignment in a product prototype.

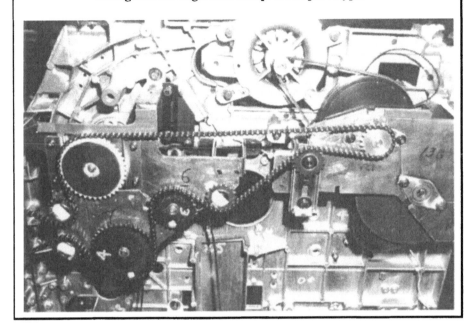

Figure 4.14   The noise spectrum of the prototype shows tonal components due to the motor and gears up to 1.2 kHz and broad band noise due to the chain and sprockets above 1.2 kHz.

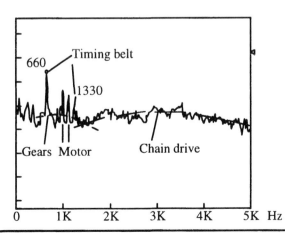

In light of the discussion above, we need to increase the chain/sprocket interface mobility in order to reduce the forces at that interface and the noise that is generated as a result. We can make that change by increasing the mobility of the chain or the sprocket(s) or both. We shall first consider the chain. Figure 4.15 shows three chains of potential use: a bicycle all-steel chain, a steel chain with links encased in plastic (a chain used in the food industry), and a "chain" made from two steel cables encased in plastic and molded into chain form. These are presented in order of increasing mobility.

Figure 4.15   Chains of varying compliance are used to confirm that interface compliance will reduce the chain-sprocket noise:

(a) a "chain" made from 2 cables encased in soft plastic with plastic cross members (not durable);

(b) a steel chain encased in a fairly rigid plastic (fairly expensive); and

(c) a conventional steel chain (a practical application, but noisy).

Figure 4.16 presents third-octave A-weighted sound power spectra for the drive system with these chains installed in sequence. It is clear that the softer cables in plastic (CinP) chain is the quietest, and the overall sound power level spectra decrease as chain mobility increases. Unfortunately the life of the CinP chain is much too short for this application, but the principle of increasing the component mobilities to reduce noise is confirmed.

Figure 4.17 shows how the mobilities of the sprockets can be increased. The drive sprockets, which must transmit torque, have their wheels and hubs separated by isolating grommets, whereas the idlers have cylindrical isolating inserts between their hubs and wheels. The results of these changes are shown in Figure 4.18. A modest reduction of noise has been achieved by these changes. If we could move the resilience closer to the sprocket/chain interface, we would likely have a greater reduction in sprocket/chain noise.

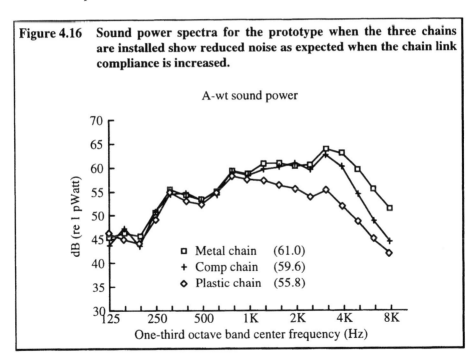

Figure 4.16   **Sound power spectra for the prototype when the three chains are installed show reduced noise as expected when the chain link compliance is increased.**

Figure 4.17   Providing compliance (or greater mobility) to the sprockets involves

(a) separating the wheel and hub of driver sprockets with resilient elements, and

(b) providing an elastomeric sleeve in the hub of idler sprockets.

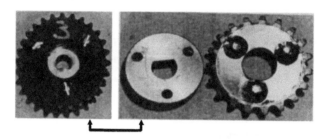

(a) assembled   (b) disassembled

Manufacturers have, of course, known that plastic gearing can be quieter, but there are trade-offs with plastic gearing that may reduce its appeal. First, the transmission error may be larger, obviating the advantage in compliance. Second, in order to maintain the life of the gearing, very rigid plastics such as forged nylon may be used. Gears that are hard enough for good life may still be inaccurate for full benefit. But the basic message is that good accuracy and high compliance will yield benefits. The benefit in plastic gearing is *not*, as often thought, a result of greater *damping* in the gearing. Damping in gearing will lead to heat rise and deterioration of plastic gears.

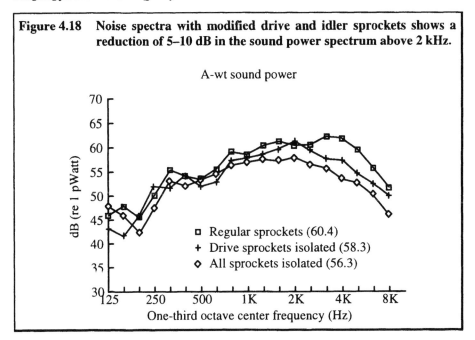

Figure 4.18   Noise spectra with modified drive and idler sprockets shows a reduction of 5–10 dB in the sound power spectrum above 2 kHz.

A-wt sound power

□ Regular sprockets (60.4)
+ Drive sprockets isolated (58.3)
◇ All sprockets isolated (56.3)

dB (re 1 pWatt)

One-third octave center frequency (Hz)

4.5

CAMS AND FOLLOWERS—IC (INTERNAL COMBUSTION) ENGINE VALVE TRAIN CAMS

Cams, four-bar linkages, and other similar mechanisms are designed to control motion in a machine. As such, their basic design is geometric, and their designers have tended to concentrate on the displacement they produce. As the speed of the machine is increased, the small irregularities in slope, curvature, or changes in curvature become more important in producing high-frequency noise energy. Cams and followers in high-speed IC engines are an example of this situation.

A diagram of the cam-driven IC engine valve (inlet or exhaust) is shown in Figure 4.19. The cam profile for this valve is shown in Figure 4.20a, and the free acceleration that this cam would produce on a follower with no mass is shown in Figure 4.20b.[7]

Figure 4.19   Diagram of valve components for an overhead valve spark ignition (SI) engine.[7]

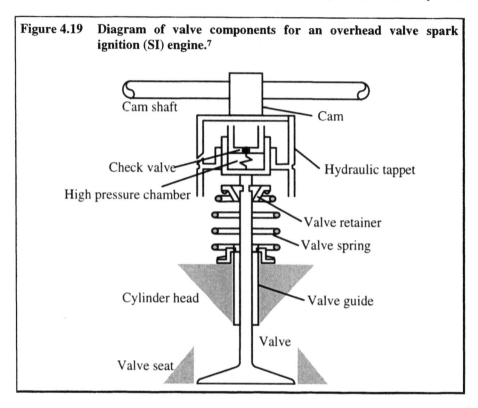

A mobility diagram of the valve and its associated components is shown in Figure 4.21. The free acceleration at the cam-follower interface is the source of noise energy, and the components define its spectrum and the path for the energy to leave the valve train. In this case, that path is the cam, camshaft, and its supporting bearings. A spectrum of the force at the cam-follower interface is shown in Figure 4.22. The frequency peak near 6 kHz is of special interest here because this noise energy is especially perceptible over the background sound from the rest of the engine.

Of course, the cam must still produce the needed valve lift height and duration, and these become constraints on the acceptable changes. A cam, redesigned to reduce this high-frequency energy, has the profile shown in Figure 4.23, and it produces the interface force spectrum shown in Figure 4.24. The small changes needed to reduce the noise energy can be accommodated without affecting the basic valve performance.

Figure 4.20 Design cam-follower displacement, velocity, acceleration, and jerk (from the geometry).[7]

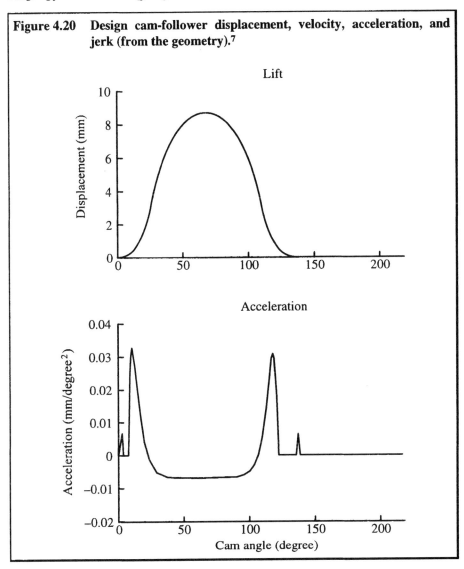

Figure 4.21   Dynamic model for the valve train, valid to about 6 kHz.[7]

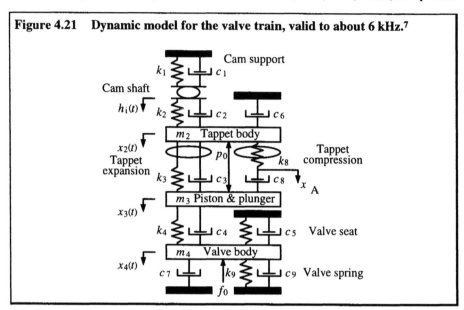

Figure 4.22   Time waveforms and spectra of cam-follower and valve head-seat forces show high frequency components due to resonances and valve bounce.[7]

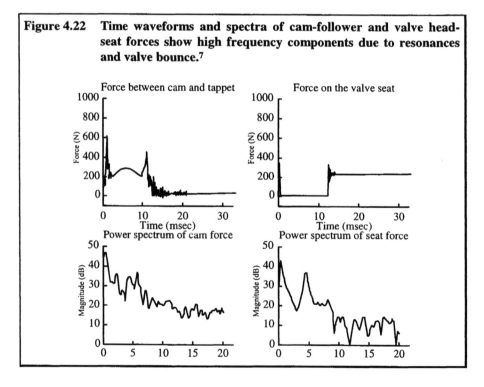

Figure 4.23   **A modified cam design has somewhat less jerk than the original.[7]**

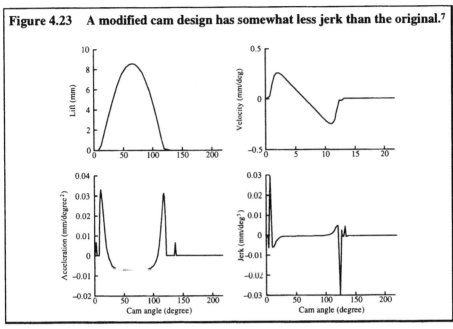

Figure 4.24   **The dynamic model can be used to estimate changes in valve train force spectra as a guide in designing a quieter valve train.[7]**

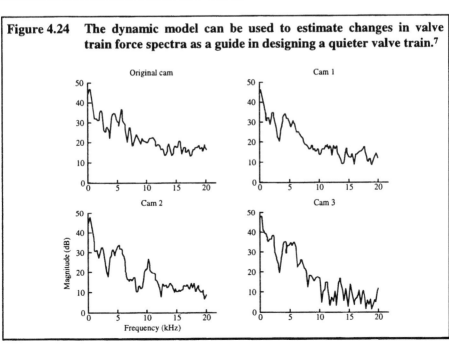

Cam and follower noise excitation are only one source of valve train noise. The other important source is the impact upon valve seating against the head. That will be dealt with in the next chapter when we discuss sources of impact.

CHAPTER 5

Sources of Prescribed Force: Electromagnetics, Airflow, Combustion, and Impact

This chapter discusses mechanisms that tend to have rather high internal mobility so that they are described primarily by the blocked force that they apply to other parts of the product. As before however, we must continue to remember that this categorization is an idealization, and there may be cases where the internal mobilities of such sources are comparable to the loads that they encounter.

5.1
ELECTROMAGNETICS AS A SOURCE: SMALL INDUCTION MOTOR EXAMPLE

Figure 5.1 is a sketch of three of the more popular designs for small motors used in a large number of products. They are split-phase induction motors, universal motors, and brushless DC motors. Induction and brushless motors are attractive because no electrical connection is needed to the rotor, and therefore brushes and a commutator are not needed. Our initial discussions will be concerned with induction motors; the others will be discussed in later sections of this chapter.[8]

Figure 5.1   Cross-sectional diagrams for three basic motor designs:

(a) induction motor,

(b) DC motor (brush type), and

(c) brushless DC motor.

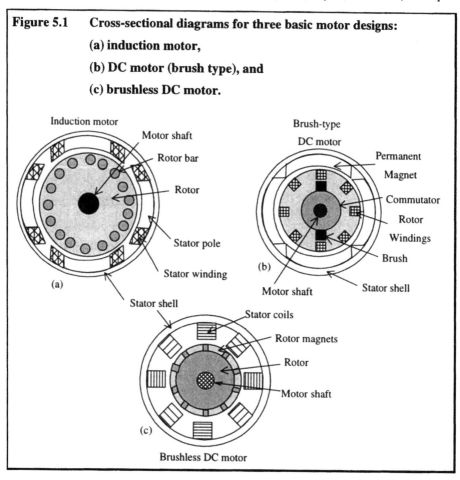

The single-phase line voltage causes two counterrotating magnetic fields in a single-phase induction motor, and the rotor tries to follow one of these. The rotating magnetic field induces currents within the electrically shorted conductors in the rotor that try to expel the magnetic field, which can only be done if the rotor speed matches the rotation of the magnetic field. If there is a load on the motor, the rotor is slowed, the magnetic field cuts through the rotor conductors (the rotor bars) producing currents in the rotor bars and the needed torque. The greater the load, the slower the rotor turns, and the greater the "slip." The slip frequency is the difference between the rotation rates of the magnetic field and the rotor.

In order to get the rotor moving from rest, a phase shift must be introduced into one pair of stator windings so that one rotating component of the magnetic field is larger than the other. This may be done with auxiliary poles having a shorted winding wrapped around them (shaded poles). It may also be done by introducing a series resistance or capacitance into one set of windings, with the capacitor being more prevalent in usage. The capacitor may be switched out after the motor starts, or it may remain connected. In the latter case, the motor is called a "permanent starting capacitor" (PSC) motor. In the following, we shall be discussing some data from a PSC motor.

There are several sources of vibration inherent in PSC motors. They include the rotational imbalance discussed previously. The best-known source is the torque fluctuation that arises from a variety of causes. When the load is light and the rotor nears synchronous speed with the field, the backward rotating field causes torque pulsations. These can produce many harmonics in the output torque, particularly if the magnetic field itself has spatial irregularities. Such irregularities occur because the windings are discrete, and the magnetic properties of the laminations in the stator and rotor are not uniform because of variations in both the geometric and material properties.

A comparison of the torque fluctuations in two motors is shown in Figure 5.2. The two motors have different degrees of "skew" in the conductors in their rotors. These measurements are of "blocked torque," i.e., as measured into a load of low (dynamic) torsional mobility. The actual torque fluctuations applied into the load, and by reaction, back into the motor will depend on the load and the internal mobility of the motor. The latter will be dominated by the inertia of the rotor itself.

Small inexpensive "shaded pole" induction motors like the one shown in Figure 5.3 are used in light-duty applications for fans in products as varied as refrigerators, humidifiers, air purifiers, and bathroom and stove vents. These motors have a noise source that is not commonly of high enough frequency in other induction motors to be audible. As the rotating magnetic field cuts through the rotor bars, if the field is not uniform, there will be a force at the slip frequency times the number of rotor bars. Normally, the slip frequency is so low that this does not produce audible sound. In a shaded-pole motor the slip frequency is a sizable fraction of the field rotation frequency, and the motors may have 30 to 60 rotor bars. Figure 5.4 shows a spectrum of sound from such a motor where the excitation produces noise due to this source at a very audible frequency of 600 Hz.

Figure 5.2   Spectra of torque fluctuations in a PSC (permanent starting capacitor) single-phase induction motor.

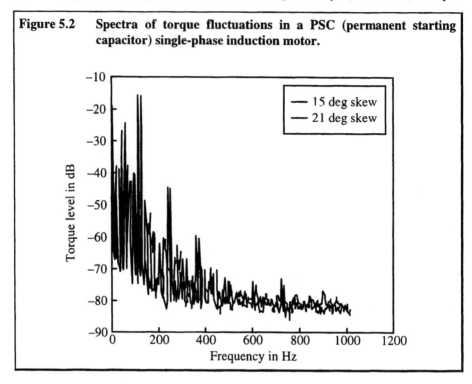

A second source of vibration can be termed *magnetic runout*, in analogy with ordinary geometric runout that occurs if a shaft is not geometrically centered in its bearings. Magnetic runout refers to a variation in the attraction force between rotor and stator as the magnetic field and the rotor spin about. If the rotor were perfectly centered in the field and the magnetic attraction were uniform about the rotor, then there would be no *net* force on the rotor. But because such perfection does not exist, there is a net force, and both rotor and stator have force excitation. Because the force is radial, this places a dynamic loading on the bearings and may cause impacts if there is clearance in the bearings. A time record showing such impacts is presented in Figure 5.5. Avoiding these impacts and the noise energy they produce is usually approached by designing the bearing supports so that this noise is attenuated, as it is very difficult to eliminate the magnetic imbalance at reasonable cost.

Figure 5.3   **(a) Cross-sectional view of a shaded-pole single-phase induction motor.**

(b) Typical torque versus speed curve.

(c) Popular applications of shaded pole motors is to fans with light load. High slip and low efficiency in these motors allow simple speed control.

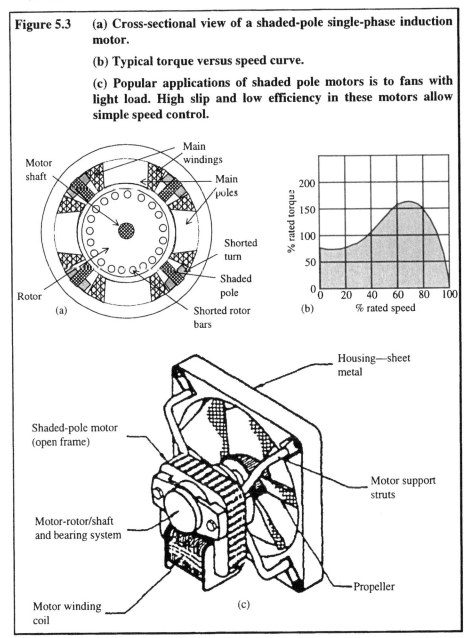

Figure 5.4   **High slip speed in shaded-pole motors and large number of rotor bars leads to audible tones, in this case at 280 Hz.**

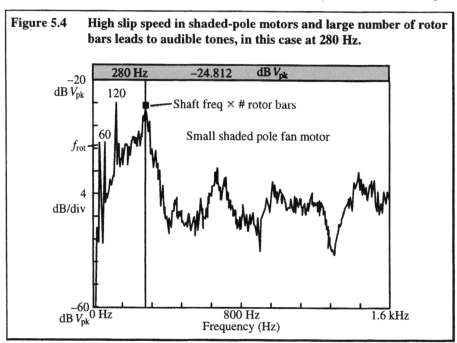

Figure 5.5   **Magnetic imbalance in motor creates radial forces and bearing impacts in the PSC single-phase motor.**

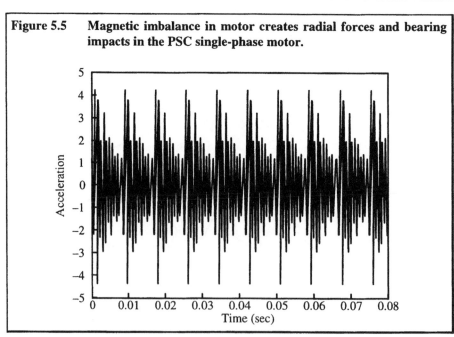

The third source of vibration is the attraction of the motor to surrounding magnetic structure, such as a steel chassis or frame supporting the motor. The desire to get more performance from small motors has led to high-flux densities in their laminations and to operating the magnetic materials at or near saturation. When this occurs, more of the flux will escape into the surrounding air, and the motor becomes an electromagnet. We refer to this source of noise as stray or fringing flux. If the power line frequency is 60 Hz, then the attractive force has a strong component at 120 Hz and harmonics. The dependence of the force at 120 Hz on distance from a steel plate is shown in Figure 5.6. The force can be reduced by either designing the motor with more lamination material, using nonmagnetic materials in the vicinity of the motor (plastic motor supports are common), or placing a layer of high permeability material (such as Mu metal) between the motor and its supports.

5.2
ELECTROMAGNETIC SOURCES; UNIVERSAL MOTORS

Typical applications for universal motors include vacuum cleaners and hand tools where high speed and moderate power are required. They are similar to the motor shown in Figure 5.1, except that they have a wound field (stator) in addition to the wound armature (rotor). Electrical contact is made with the armature windings through a commutator and brushes. Despite their name, universal motors are almost always operated on AC power. Although they will run on DC, the DC motor with commutator and brushes almost always has a permanent magnet stator, as shown in Figure 5.1. The speed of a universal motor is only limited by the load and internal losses. Typical speeds are 12000 to 30000 rpm (200–500 Hz), so that the forces due to mechanical imbalance and magnetic runout at rotor speed produce vibrations that are directly audible.

The noise spectrum of a universal motor used in a vacuum cleaner, shown in Figure 5.7, reveals several important points about the sound from universal motors. We note there are a series of tones on a varying background. The motor is turning at 24600 rpm (410 Hz), and we see a 410 Hz line and several of its harmonics. The large fundamental is probably mass imbalance, but as discussed in Chapter 4, we would normally expect weak harmonics for a rotating mass imbalance. A rotating off center mass theoretically produces imbalance force only at the frequency of rotation, but practically, rotating imbalance force has harmonics that drop off at 20–30 dB per harmonic. The harmonics in this case do not drop off and some even rise.

Figure 5.6   Leakage or fringing flux from a motor results in an attraction force on a steel chassis.

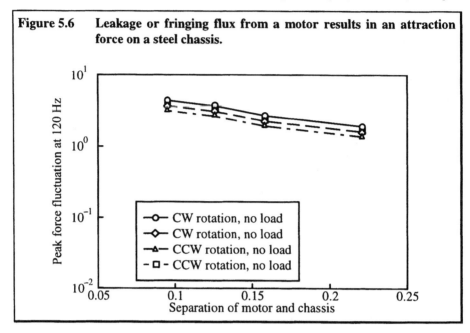

Figure 5.7   Spectrum of sound radiated by a universal motor shows effects of mechanical imbalance, magnetic runout, and structural resonances.

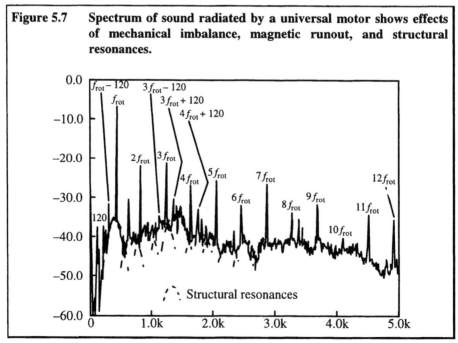

The sketch in Figure 5.8 shows a rotating armature in a magnetic field. The magnetic flux in the air gap determines the attractive force between rotor and stator, and that flux in turn depends on the reluctance of the magnetic circuit. In most motors, that reluctance is dominated by the air gap. As shown, when the rotor turns, sometimes there are three rotor slots in the circuit, sometimes four, causing the reluctance, the magnetic flux, and the attractive force to vary. A typical universal motor will have 10–20 armature slots, leading to a slot passage frequency of 10–20 times the rotation frequency. In the motor in Figure 5.7, there are ten slots, and the slot passage frequency is 4.1 kHz.

Magnetic imbalance was discussed in the previous section, and it occurs in universal motors also. It shows up in Figure 5.7 in the rotation frequency harmonics and their 120 Hz modulation sidebands. If the stator field were from a permanent magnet, magnetic imbalance would produce only harmonics of the rotation frequency, but because the magnetic attraction is switched on and off 120 times a second, these forces are modulated, and the motor speed harmonics have sidebands separated by 120 Hz from them.

Finally, the underlying broadband background in the spectrum of Figure 5.7 shows a series of "humps" that are caused by the structural resonances of the motor. We see that unfortunately the radiation at 410 Hz is sitting right on top of a resonance that is increasing the sound radiation by about 10 dB. Obviously, a way to reduce that tonal component in this case is to move the resonance away from the operating speed of the vacuum cleaner. This would be done by structural changes to the motor.

5.3
ELECTROMAGNETIC SOURCES; DC BRUSHLESS MOTORS

DC brushless motors and variable frequency induction motors are becoming more popular for applications where one would like to control motor speed and direction and not have to worry about making electrical connections to a rotor. These motors are driven by a multiphase electrical supply of controlled frequency. In the brushless DC motor (DC because DC supplies the power source, but the excitation is pulsed.), the rotor is magnetized, and the rotation is synchronous with the excitation. Such motors are useful when a certain speed is to be maintained, as in a computer disc drive or computer projector color wheel.

Figure 5.8 **As the rotor moves, the magnetic reluctance changes leading to a variation in the attraction force between stator and rotor.**

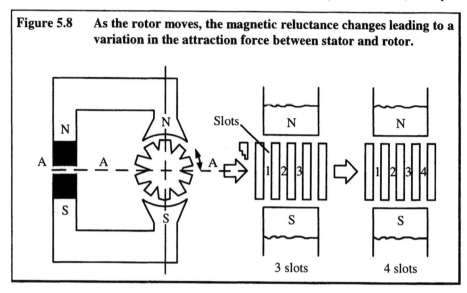

The voltage supplied to such motors is far from sinusoidal. Figure 5.9a shows the voltage waveform applied to a variable-speed induction motor using pulse width modulation to control the motor power. Figure 5.9b shows the current waveform, which is much more sinusoidal, showing the effect of the stator inductive reactance and the inertia of the rotor in smoothing the current (the rotor mass looks like a large capacitor when reflected to the electrical side of the electromechanical transformation).

A sketch of a brushless DC motor in a disc drive application is shown in Figure 5.10. At the operating speed of the motor, the fundamental frequency of the voltage excitation is 2.72 kHz, and that frequency shows up strongly in the sound radiation as shown in Figure 5.11. Tap resonance tests of the motor structure, shown in Figure 5.12, do not show resonances of the motor structure very near this frequency, although it is quite common for the voltage waveform to have a fundamental or harmonic that will excite a structural resonance of the motor or attachment (like a computer disc platen or color wheel).

We discussed the use of perforations in reducing radiated sound in Chapter 3, and we shall have more to say on the subject in Chapter 9, but we can show here how the idea can be applied to a motor. If we perforate the two halves of the motor shell as shown in Figure 5.13, the sound radiation is as shown in Figure 5.11. The reduction in radiated sound at 2.72 kHz is about 20 dB, which is an usually large benefit to gain from such a simple step. In this case, the perforation is acceptable because in the application, dust or other potential contaminants are excluded from the motor environment. If dust were a consideration, the holes would have to be covered with a lightweight screen or gauze.

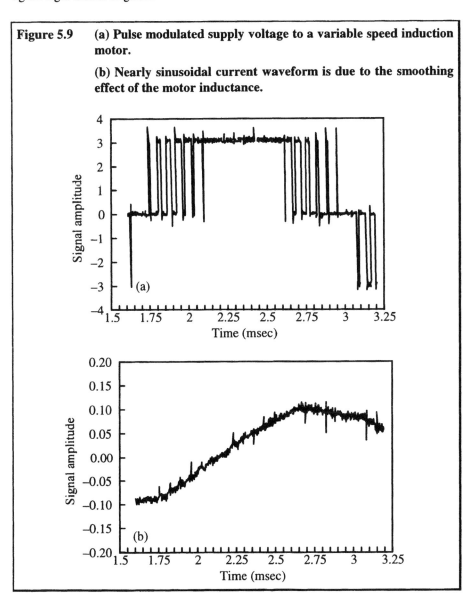

Figure 5.9 (a) **Pulse modulated supply voltage to a variable speed induction motor.**

(b) **Nearly sinusoidal current waveform is due to the smoothing effect of the motor inductance.**

Figure 5.10   Exploded view of computer disc drive that incorporates a synchronous brushless DC motor.

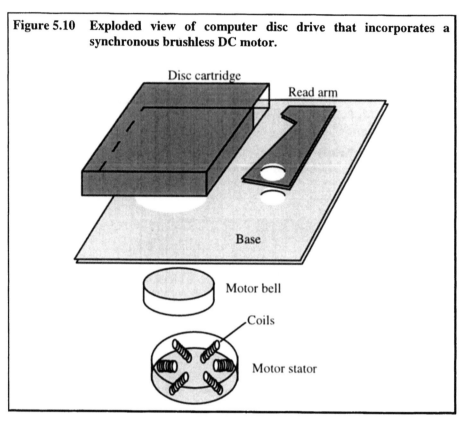

Figure 5.11   Pattern of perforations in bell and rotor of motor to produce the reduction in sound radiation.

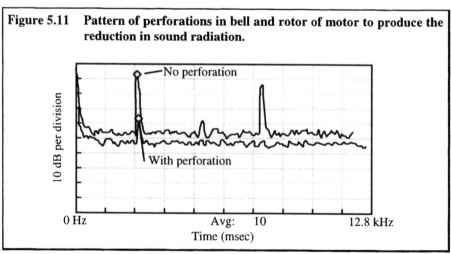

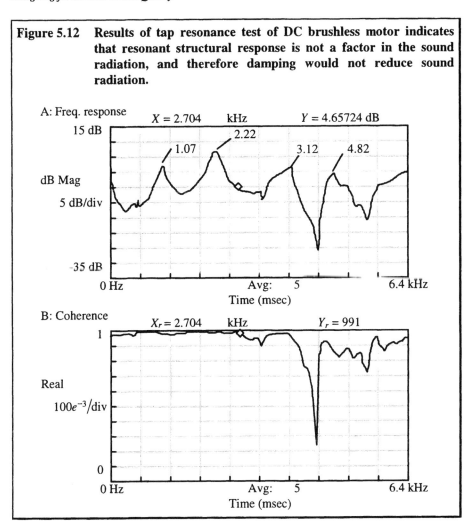

Figure 5.12 Results of tap resonance test of DC brushless motor indicates that resonant structural response is not a factor in the sound radiation, and therefore damping would not reduce sound radiation.

5.4
AIRFLOW SOURCES; HEAD LOSS AND GAIN

In Chapter 8 we discuss various mechanisms for sound generation, including the sound produced by an object exerting a dynamic force on the air. In products that employ airflow for cooling, there is a source of head (pressure) rise, a fan or blower, and head loss in the airflow passages. Although most attention is usually placed on the fan as the noise source—and properly so—the head loss is also a source of broadband noise that represents an unavoidable lower limit to the final noise level of the product after all other sources are eliminated, unless the overall resistance to flow through the product can be reduced.

Head loss in the air passages through the product generally result from turbulent drag, and the forces resulting from the shedding of turbulence are random. These dynamic forces are approximately 30 percent of the average drag force, and they have a time scale related to the average flow velocity and the dimensions of the flow passages. Although it is not easy to give precise values to all these parameters, the spectral shape of the noise is rather smooth and somewhat insensitive to details. A simple formula for the overall radiated sound power is given by

$$\Pi_{rad} \approx 2 \times 10^{-6} Q^6 / d_h^{10} \tag{5.1}$$

where Q is the volume flow through the unit in CFM and d_h is the hydraulic diameter of the flow passages (in cm). Typical values of Q are 1–10 CFM and 0.1–1 cm for d_h, leading to power levels in the range of 30–70 dB re 1 pW (1 pW $= 10^{-12}$ watts). At the upper end of this range, products designed for a quiet environment can be too noisy. The third-octave band frequency spectrum of this noise is often represented as shown in Figure 5.14, a "haystack" spectrum. The normalizing frequency is the Strouhal frequency,

$$f_S = 30Q / d_h^2 \tag{5.2}$$

Figure 5.13   **Sound radiated by DC brushless motor at electrical drive frequency and harmonics is substantially reduced by perforations (note 5 dB offset in data for visualization).**

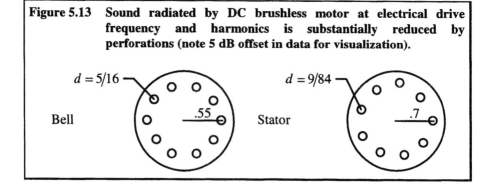

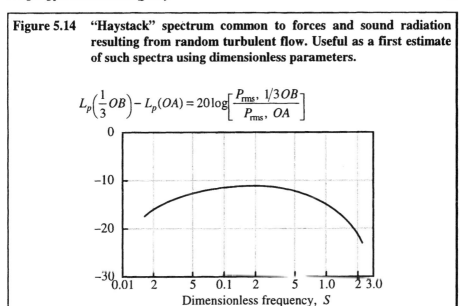

Figure 5.14   **"Haystack" spectrum common to forces and sound radiation resulting from random turbulent flow. Useful as a first estimate of such spectra using dimensionless parameters.**

$$L_p\left(\frac{1}{3}OB\right) - L_p(OA) = 20\log\left[\frac{P_{rms},\ 1/3\,OB}{P_{rms},\ OA}\right]$$

Although a fan operating at an appropriate load point will still make noise, the fan will usually be noisier if it is not properly loaded. Therefore, the first step in controlling fan noise is to match the desired flow volume and head rise to the fan. It is surprising how often this is not done, in part because the proper data on the fan characteristics is not readily available, or because the supplier cannot provide a fan with the desired characteristics. A typical head versus flow fan curve is shown in Figure 5.15. The range of operating points where the fan is efficient is indicated in the figure.

Once the proper operating point is chosen, the fan will still have two major sources of dynamic force between its surfaces and the air and therefore of sound. One source is random noise due to turbulent flow at the fan inlet. When this flow strikes the blade, a force results depending on the sensitivity of the lifting force to angle of attack. A formula for the sound power generated by this source is given by

$$\Pi_{rad} \approx 6 \times 10^{-12} Q^6 / d_h^{10} \alpha^4 \tag{5.3}$$

where α is the "angle of attack" of the rotating blade in radians as it meets the incoming airflow. The formula assumes fluctuations in the incoming airstream of about 10 percent of the mean flow velocity. Because this source is directly dependent on the level of turbulence in the inlet flow, the object of design is to keep the inflow turbulence as low as possible by keeping the inlet path as unobstructed as possible and any irregularities in the upstream or downstream passages at least 1 fan radius away from the fan.

The second source due to the fan is tonal, usually at the blade passage rate, caused by spatially fixed irregularities that produce a wake in the inlet flow (guide vanes, fan support struts) or at the outlet (scroll diffuser cutoff). Figure 5.16 shows how rapidly the force on a fan blade changes in both magnitude and direction as the blade passes by the scroll diffuser cutoff. There is also a similar force fluctuation on the cutoff that is not shown. These dynamic forces can produce strong tones at the blade passage frequency.

The usual "fix" to the cutoff-induced tone is to round the cutoff and move it as far away from the blades as possible, which can still result in an effective fan if properly done. Figure 5.17 shows the fan and scroll from a consumer vacuum cleaner for which that has been done.

Figure 5.15   Performance curves for centrifugal fans with forward and backward curved blades.

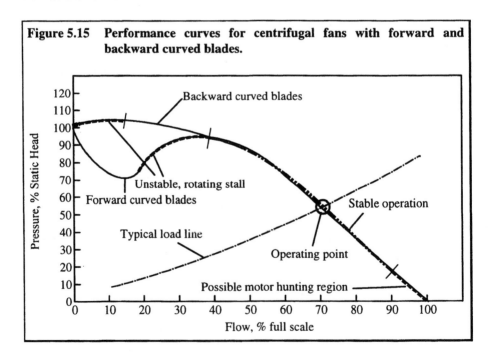

Figure 5.16 Centrifugal fan with volute (scroll) diffuser has rapidly changing forces on the blades and cutoff as the blade passes the cutoff, producing a tone in the sound at the blade passage frequency.

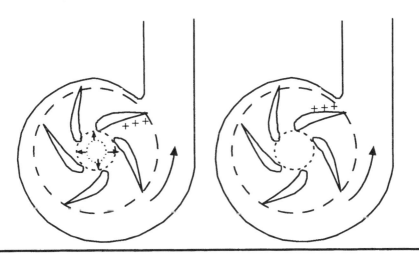

Figure 5.17 Commercial centrifugal fan with airfoil-shaped blades and variable depth volute produces only a weak blade passage tone.

5.5

COMBUSTION PRESSURE IN IC ENGINES

The transient pressures within the combustion chamber of an IC engine like that shown in Figure 5.18 push upward on the head and downward on the piston to create noise energy that may escape to engine surfaces and radiate sound. The high-frequency content of the combustion pressure in a spark ignition IC engine is usually so weak that this sound is inaudible, but it is a factor in the sound of a diesel engine.

A cross section of the combustion chamber of a direct injection diesel is shown in Figure 5.19. Figure 5.20 shows an overlay of the cylinder pressure in such an engine with and without the cylinder firing. The onset of combustion shows up as a slight increase in the rate of pressure increase. When the engine fires, the peak pressure rise in the cylinders is only 20–30 percent greater than that due to compression alone if it doesn't fire. But the increase in frequency components in the audible frequency range, shown in Figure 5.21, is due only to the break in slope of the pressure-time curve when combustion begins, a feature that is nearly invisible unless one is looking for it.

Figure 5.18   **Cross-section sketch of a four-cylinder diesel engine shows its major structural components.**

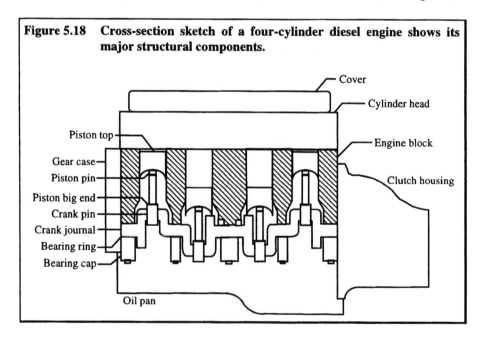

Figure 5.19   **Configuration of the combustion chamber of a direct injection (DI) diesel engine.**

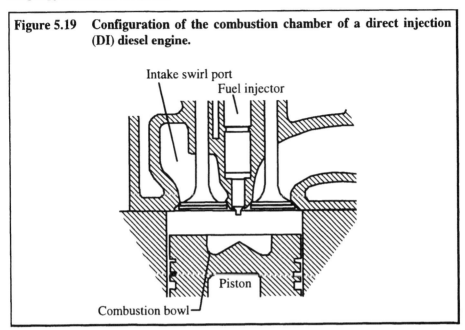

Figure 5.20   **Typical cylinder pressure curve on the compression-combustion strokes for a DI diesel engine with and without combustion.**

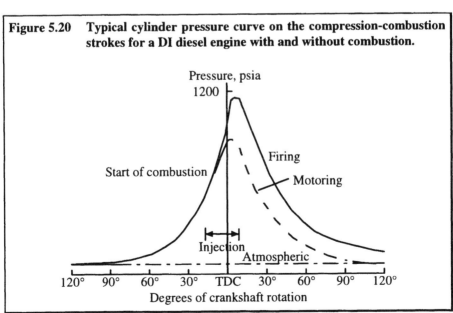

Figure 5.21   Spectrum of cylinder pressure for a DI diesel engine with and without combustion shows that combustion produces most of the noise energy in the audible frequency range.

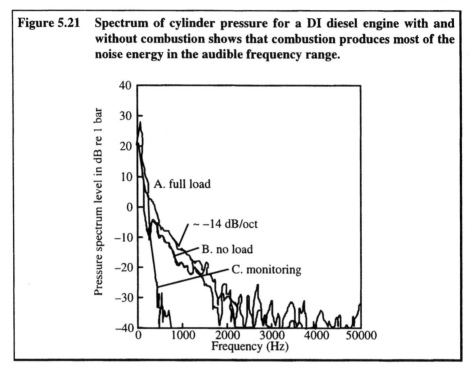

There are many factors that determine the high-frequency content of diesel combustion pressures and whether or not they contribute in any important way to the radiated sound. If the engine is turbocharged, the combustion pressure tends to be smoother, reducing the noise energy. An indirect injection engine, with the combustion chamber shown in Figure 5.22, has a different combustion pressure waveform as shown in Figure 5.23, but the general spectral character is similar to that shown in Figure 5.21.

5.6
IMPACT AS A SOURCE; PISTON SLAP IN IC ENGINES

Impacts occur in products both by accident and design. Loose parts, joint clearances, and vibrating components placed close to other all produce impacts that are very audible and undesirable. Mechanisms such as crank-sliders, four-bar linkages, and cams and followers that are designed into products can also impact at places in their cycles when the interaction force vanishes so they can float free, only to be brought together again a moment later with a bang.

Figure 5.22 Configuration of the combustion chamber of an indirect injection (IDI) diesel engine showing its combustion prechamber.

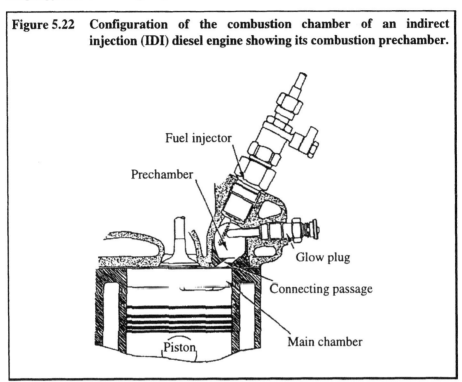

Figure 5.23 Cylinder pressure of an IDI engine shows late onset of combustion pressure as compared to the DI engine.

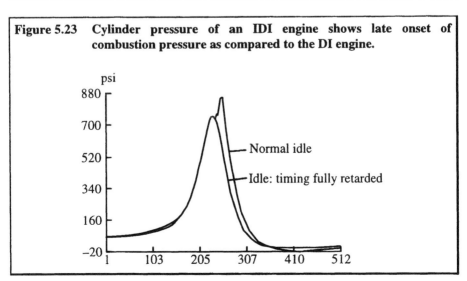

A classic problem is the impact of a resilient mass impacting a rigid surface shown in Figure 5.24.[9] The mass approaches the surface with velocity v_0 and rebounds from it, creating a pulse of force on the surface. During contact, the compression of the ball is modeled as a spring with stiffness K (the spring represents the storage of potential energy in the ball). During contact, therefore, the dynamics are those of a mass bouncing on a spring. The velocity, displacement, and spring force are graphed in Figure 5.25. Because the mass is not attached to the surface, when the force vanishes and the velocity is negative the mass leaves the surface and the force remains zero. The waveform of the force is the half-sine shown with a contact time and peak force given by

$$T_c = \pi(m/K)^{1/2}; \quad f_{max} = v_0(m/K)^{1/2}. \tag{5.4}$$

The frequency spectrum of this pulse is

$$E(\omega) = \frac{(2mv_0)^2}{2\pi(1 - \omega^2/\omega_{res}^2)^2} \cos^2 \frac{\pi\omega}{2\omega_{res}} \tag{5.5}$$

and is graphed in Figure 5.26. This is the approximate spectrum produced by an impact hammer, a device used in vibration testing. We note that the corner frequency is $1.5/T_c$, so we can concentrate the force applied to a structure to lower frequencies by using a softer tip on the hammer that increases the contact time. Steel, plastic, and rubber tips are available for impact hammers so that some control of the excitation frequency spectrum is possible.

Figure 5.24   **Impact of a ball against a rigid surface shows how kinetic energy (stored in the mass) is converted to potential energy (modeled as a spring) and converted back again to kinetic energy on the rebound.**

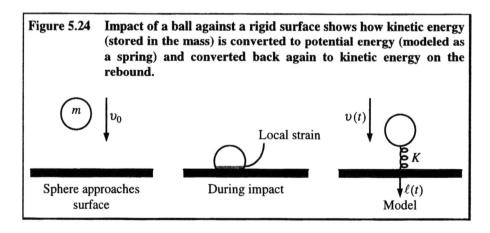

Sphere approaches
surface

During impact

Model

Figure 5.25 Waveforms of velocity and displacement of ball center of gravity (CG) determine the force applied to the surface as a half-sine pulse.

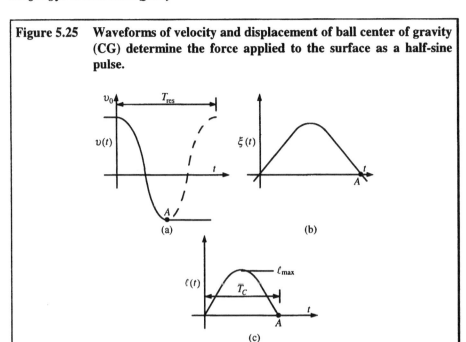

Figure 5.26 Low-frequency part of spectrum of impact force depends on momentum (impulse) imparted to the surface, whereas the type of force discontinuity determines the high-frequency behavior. The corner frequency depends on the contact time.

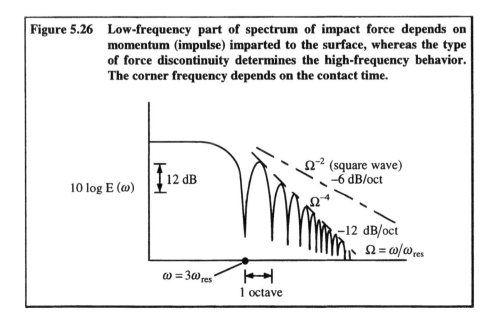

We have seen how the mechanical properties of the impactor affect the force spectrum; we now look at the example of piston slap in an IC engine where both impactor and impactee combine to define the force spectrum. During the two complete revolutions of a four-stroke engine, the piston moves from one side of the cylinder bore to the other six times, but the motion from the antithrust to the thrust side immediately after top dead center (ATDC) on the compression stroke usually produces the strongest impact and audible sound, particularly in a diesel engine. The sketch in Figure 5.27 shows how the pressure above the piston, combined with the changing angle of the connecting rod causes the piston to slide across the free clearance in the bore and impact the thrust side of the bore. The higher cylinder pressure and greater clearance in a diesel engine are responsible for the stronger impact associated with such engines.

The piston approaches the cylinder wall at velocity v_0. When it contacts the wall, the combination of cylinder wall and piston begins to vibrate with initial velocity v_0. The dynamics of the cylinder wall are determined by measuring the drive point mobility of the bore wall where the impact occurs and are shown in Figure 5.28.[10] This data is modeled by a simple stiffness of value $K_w = 1.58 \times 10^8$ N/m. The drive point mobility of the piston is shown in Figure 5.29 and can be modeled as two masses and two springs. The masses correspond roughly to the mass of the piston and to that of the upper part of the connecting rod. The stiffnesses correspond to those of the piston skirt and of the wrist pin between the piston and con rod. The values are shown in Figure 5.30.

Figure 5.27   Piston slap just after top dead center (ATDC) on the power stroke is a source of noise in diesel engines.[9]

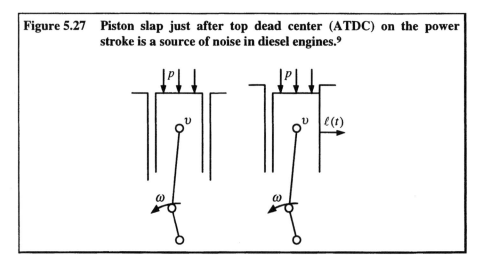

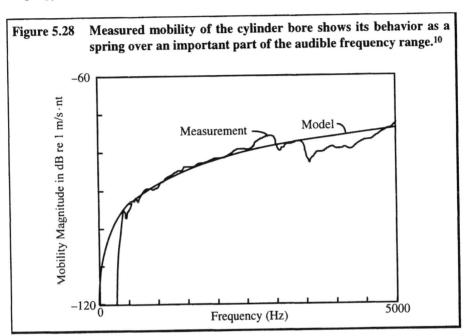

Figure 5.28   **Measured mobility of the cylinder bore shows its behavior as a spring over an important part of the audible frequency range.[10]**

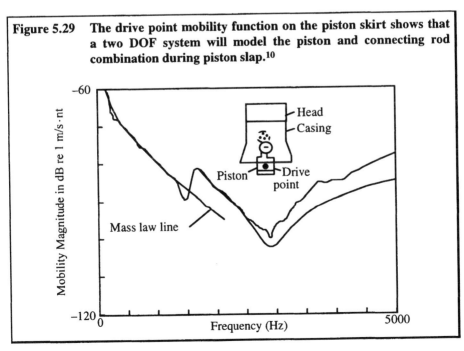

Figure 5.29   **The drive point mobility function on the piston skirt shows that a two DOF system will model the piston and connecting rod combination during piston slap.[10]**

Figure 5.30   Combining models for the cylinder bore, piston, and connecting rod provides a dynamic model for piston slap.[10]

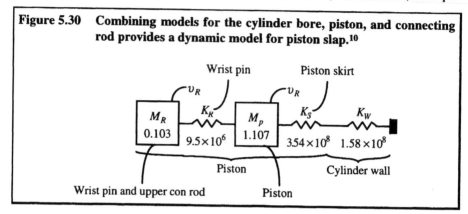

The computed spectrum of cylinder wall vibration from the model shown in Figure 5.30 is presented in Figure 5.31. This engine was rotated using an electric motor with piston and con rod present only in one cylinder. If the valves in that cylinder remain closed and the cylinder is pressurized, the pressure in the cylinder is very close to the combustion pressure, simulating piston slap when combustion is present.

Figure 5.31   Comparison of the spectrum for measured piston slap-induced vibration and the spectrum calculated using the model in Figure 5.30.[10]

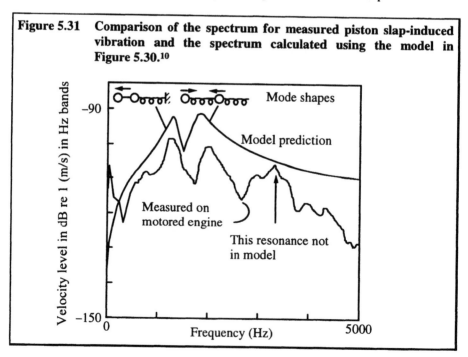

In this situation, the piston slap impact was measured to be as in Figure 5.32. The spectrum of this pulse is also shown in Figure 5.31. We see a good correspondence between the two lower spectral peaks at about 1300 and 2200 Hz, but the experimental peak at 3300 Hz is not shown, probably because that resonance has not been modeled. It may correspond to a rocking mode on the piston against the bore wall, because the piston is designed to rotate about the wrist pin as it traverses the clearance. We also note that the theoretical calculation overpredicts the value of the spectrum by about 8–10 dB, probably because the theory assumes free flight across the clearance space, whereas in reality, the translation of the piston will be impeded by ring friction and viscous drag from the oil film. In Chapter 7, we will see how the various sources we have discussed for a diesel engine will combine to produce an overall vibration level for the engine casing.

Figure 5.32 **Measured vibration (acceleration) of the cylinder bore due to piston slap. Because the bore acts as a spring, when acceleration is converted to displacement the waveform will indicate the force between piston and bore during piston slap.[10]**

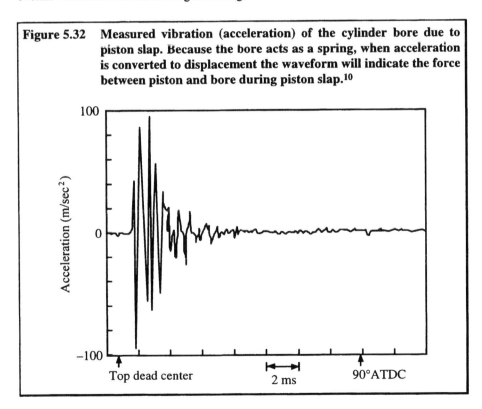

We have noted that the piston slap just after top down center (TDC) on the compression stroke is usually the strongest impact, and the impact is audible in diesel engines. It is usually inaudible in spark ignition (SI) engines, particularly if they are water cooled because the sound is masked by other noise sources in the engine. Air-cooled SI engines may produce audible piston slap because the water jacket that isolates the sound is not present and because air-cooled engines normally have greater piston to bore clearance. In very rare cases, there may be piston slap sound that arises from one of the other five piston motions from one side of the bore to the other, but some problem in the engine is likely causing that to happen.

5.7
IMPACTS DUE TO LOOSENESS; RATTLING

In Chapter 2 we discussed sounds from products that are expected and unexpected. Rattles, although not unusual in products, are unexpected because we expect products to be designed so that rattles are avoided. Most automobile producers have special task forces and research studies on "squeak and rattle," as such sounds are highly undesirable in the product.

Rattles result when there is clearance between structural components and accelerations are sufficient to produce impacts. The usual solution is to try to join components together (not necessarily rigidly) so they remain in contact.

CHAPTER **6**

Structural Response to Excitation; Single DOF

This chapter shows us the power of a very simple structural model, the single-dof resonator. The mass sitting on a spring is the introductory topic in books on dynamics and vibrations. And yet, it pervades ideas and applications in both structural response and sound radiation in very practical ways. Here we use the model to understand vibration isolation concepts and the dynamics of some complicated structures.

6.1
IMPORTANCE OF THE SINGLE DOF—IN ITS OWN RIGHT AND AS A MODEL

The single-dof resonator, modeled as a mass on a spring, is the simplest resonant dynamic system we might imagine, but it nevertheless has an extensive range of applications. Some systems are extremely well modeled by such an arrangement. The sewing machine sitting in a cabinet is one, as we shall show. A machine sitting on isolators is another, although the single-dof model must be regarded as the "zero order" approximation to how isolation systems actually behave. Finally, even a very general model of a vibrating system that describes the dynamics in terms of resonant modes makes use of the single-dof system as a model for each resonant mode.

The model for a single-dof resonator and its mobility diagram are shown in Figure 6.1.[11] The three elements have a common velocity so the mobilities of the mass $Y_m = 1/j\omega m$, stiffness $Y_K = j\omega/K$, and damping $Y_R = 1/R$ are combined in parallel (because velocity is the "drop" variable) to obtain the drive point mobility

$$Y_{dp} = \frac{1}{(1/Y_m + 1/Y_K + 1/Y_R)} = \frac{j\omega}{m(\omega_0^2 - \omega^2 + j\omega\omega_0\eta)} \tag{6.1}$$

where $\omega_0 = (K/m)^{1/2}$ is the undamped natural resonance frequency and $\eta = R/\omega_0 m$ is the damping loss factor (the reciprocal of the quality factor Q). The magnitude of this function is graphed in Figure 6.2. The low-frequency mobility is $Y_{dp} \approx Y_K$ and the high-frequency mobility is $Y_{dp} \approx Y_m$, so we speak of the mobility being stiffness controlled at low frequencies and mass controlled at high frequencies.

At resonance, $\omega \approx \omega_0$, we have $Y_{dp} \approx Y_R = 1/\omega_0\eta\, m$ and the response is damping controlled. Because the power dissipated by the resonator is

$$\Pi_{\text{diss}} = R\langle v^2 \rangle = \langle f^2 \rangle R |Y_{dp}|^2,$$

if we prescribe f then the frequency dependence of $|Y_{dp}|^2$ determines the power dissipated and power absorbed by the resonator from the source. If the force is random noise, then the curve in Figure 6.2 can be replaced by a uniform response $1/R^2$ over a bandwidth $\Delta_n = \pi\omega_0\eta/2$, the noise bandwidth. We often find it useful to think of the resonator responding as a mass, spring, or damper in the three frequency ranges indicated in Figure 6.2.

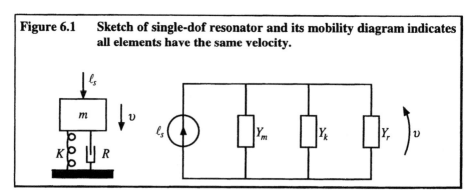

Figure 6.1   **Sketch of single-dof resonator and its mobility diagram indicates all elements have the same velocity.**

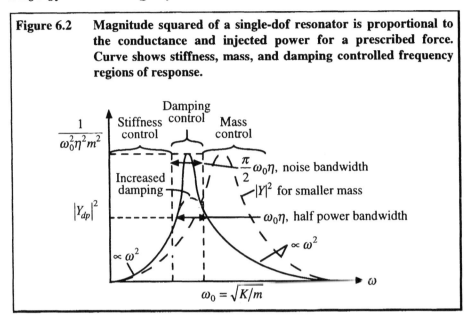

Figure 6.2   **Magnitude squared of a single-dof resonator is proportional to the conductance and injected power for a prescribed force. Curve shows stiffness, mass, and damping controlled frequency regions of response.**

In the damping controlled frequency range, the kinetic and potential energies of vibration are approximately equal. During free resonant vibration or under excitation by a random source, the damping controlled response dominates the motion, and therefore we can assume equality of these two forms of energy. Because the motion of the resonator is usually the more easily measured, we tend to relate energy to motion, and because the kinetic and potential energies are equal, we get the useful approximations $E_{\text{vib}} = M\langle v^2 \rangle = M\omega_r^2 \langle y^2 \rangle = M\langle a^2 \rangle / \omega_r^2$. The energy of vibration can therefore be determined by the ms displacement, velocity, or acceleration of the resonator mass.

6.2
RESONATOR RESPONSE TO IMBALANCE; SEWING MACHINE AND CABINET EXAMPLE

If the operating frequency of a machine should coincide with a combined resonance of the machine and its supporting structure, then unacceptable levels of vibration can easily result. An example is the sewing machine sitting on a sewing table, sketched in Figure 6.3. The imbalance force that was determined in Chapter 4 and presented in

Figure 4.6 becomes the driving force for a mode of vibration in which the machine and the cabinet both participate. A mobility diagram for the combined system is shown in Figure 6.4. The element M_m is the machine mass and represents the kinetic energy in the machine itself. The quantity ψ_s is a "mode shape factor," which in this case is simply the cosine of the angle between the direction of the force and the vibratory motion of the machine. If, as we discussed in Chapter 4, by a balancing mass we could modify the imbalance force *direction* so it were orthogonal to the direction of vibration, the factor ψ_s would vanish and the resonance would not be excited.

Figure 6.3   **Sewing machine sitting on a sewing cabinet shelf can be modeled as a single-dof resonator for the "bouncing mode."**

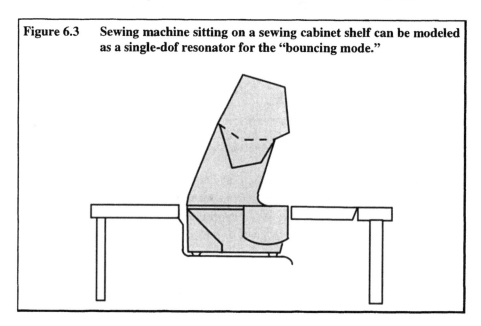

Figure 6.4   **Mobility diagram for the sewing machine shows the imbalance as the excitation, with the machine storing kinetic energy and the cabinet storing both potential and kinetic energy.**

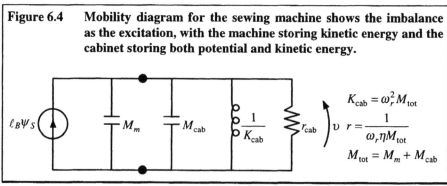

$$K_{cab} = \omega_r^2 M_{tot}$$
$$r = \frac{1}{\omega_r \eta M_{tot}}$$
$$M_{tot} = M_m + M_{cab}$$

The quantity M_{cab} represents the kinetic energy in the cabinet structure and K_{cab} represents the potential energy of vibration. It is assumed here that all of the potential energy and all of the damping represented by the conductance r_{cab} are contributed by the cabinet. If the system resonance $\omega_r = (K_{cab}/M_{tot})^{1/2}$, where $M_{tot} = M_m + M_{cab}$ is the total effective system mass, coincides with the operating frequency of the machine, or a low-order harmonic, then excessive vibration may occur.

The vibration of a sewing machine might be excessive and unacceptable for various reasons. One might actually see the vibration as a blurring of the edge of the machine or table, which is directly related to the displacement of the machine. Pins and other objects placed on the table or machine might rattle, which is a separate effect that is caused by acceleration. Or we might feel the motion of the machine, which is more complicated, but has been found to be primarily related to vibratory velocity. Figure 6.5 presents a chart that is useful in presenting such mixed criteria. If we decide that edge blur will be noticeable if the peak-to-peak displacement exceeds 0.5 mm, then the area of the chart below the 0.5 mm displacement line is acceptable. If we expect parts might rattle or move around if the peak-to-peak acceleration is too high, then the vibration must also be below the 0.5 g line.

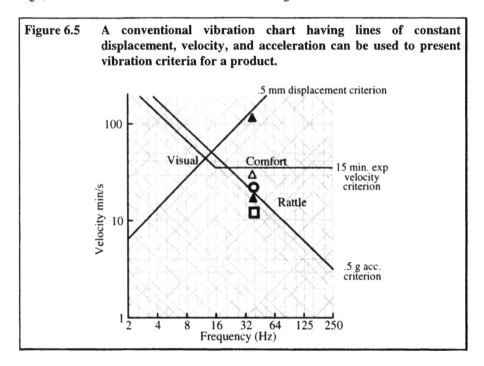

Figure 6.5   A conventional vibration chart having lines of constant displacement, velocity, and acceleration can be used to present vibration criteria for a product.

The 15-minute tactile comfort criterion (assumes 15-minute stretches of consistent sewing) is a combination of a velocity limit above and an acceleration limit below 16 Hz. We can see that the comfort criterion has little additional effect on limiting vibrations compared to the visual and rattle criteria. On the chart we have also shown some vibration levels due to the 40 Hz second harmonic of the operating frequency of the machine at 1200 stitches/minute or 20 Hz. We see that the machine in the cabinet slightly exceeds the chosen rattle criterion. Note that the vibration when the machine is in the cabinet is about twice as large (6 dB greater) than when it is freely suspended, that is, the cabinet-machine resonance *amplifies* the free vibration, theoretically by the quality factor $Q = 1/\eta$ (see Equation 6.1).

6.3
ISOLATION SYSTEM MODELED BY THE SINGLE-DOF RESONATOR

When a machine is placed on isolators, the object of isolation is to reduce the force transmitted to the base or foundation as in Figure 6.6. One way to think of the situation is to allow the blocked force from the internal activities in the machine to accelerate mass, because mass is the only element that can attenuate or absorb force (massless dampers and springs transmit force from one end to the other without attenuation). Therefore, we want the operating frequency of the machine to be above the resonance frequency so that the system is mass controlled. The model of the isolated machine is shown in Figure 6.1, and the *displacement y* as a function of the frequency of excitation is graphed in Figure 6.7a, where the parameter on the curves is $\zeta = \eta/2$, expressed as the *percentage or fraction of critical damping*. Because Equation 6.1 determines the velocity v, we can calculate the transmitted force as $f_{trans} = Rv + Ky$ to obtain the graph in Figure 6.7b. At about one-half octave above the mounted resonance frequency we begin to obtain the desired attenuation of force, but we also see that damping in the mount increases the transmitted force. But some damping is desirable in order to limit the vibrations as the machine coasts to a rest and the rotating frequency passes through the resonance. Therefore the amount of damping in the mount must be chosen with care.

This analysis predicts a constantly decreasing force at any frequency as the frequency of the mounted resonance is reduced, but that ignores the fact that the isolator may have to become unreasonably soft. The isolator resonance frequency can be conveniently determined from the static displacement of the machine when placed on the isolators $\delta_{\text{stat}} = mg/K$;

$$f_r = \frac{1}{2\pi}\left(\frac{K}{m}\right)^{1/2} = \frac{1}{2\pi}\left(\frac{Kg}{mg}\right)^{1/2} \approx \frac{100}{\delta_{\text{stat}}^{1/2}(\text{mils})} \tag{6.2}$$

Because there is a necessity to lower f_r in order to achieve more isolation, the static displacement will have to be increased. For example, if the static displacement is 4000 mils (4 in.) then $f_r = 1.6$ Hz. This is about the lowest resonance frequency one can achieve using a passive mounting system, as it is difficult to implement isolation systems with springs long enough and soft enough to achieve such large values of static deflection.

Longer and softer springs are also more likely to have surge (internal) resonances that increase the transmitted forces. Figure 6.8 graphs the effects of isolator mass and damping on this transmissibility. As a practical matter, surge resonances at 1–2 decades above the design resonance limit the reduction of transmitted force to 20–30 dB.

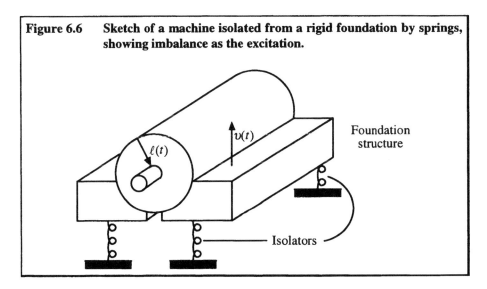

Figure 6.6   **Sketch of a machine isolated from a rigid foundation by springs, showing imbalance as the excitation.**

Figure 6.7   **Displacement amplitude and force transmission to a foundation for various values of damping. The force transmissibility shows that damping has a beneficial effect at the bounce resonance frequency, but increases the transmitted force at higher frequencies.**

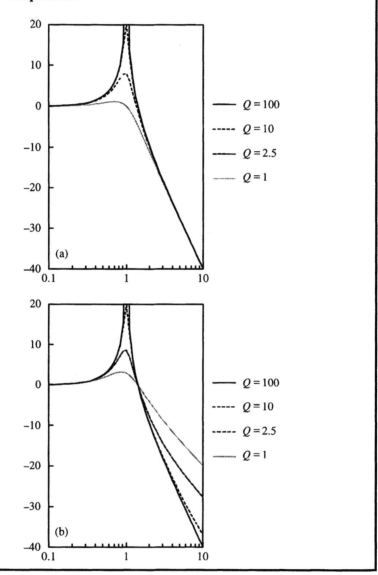

Figure 6.8 Force transmissibility of an isolator showing the effects of surge, or internal resonances of the isolator, in increasing the high-frequency force transmission.

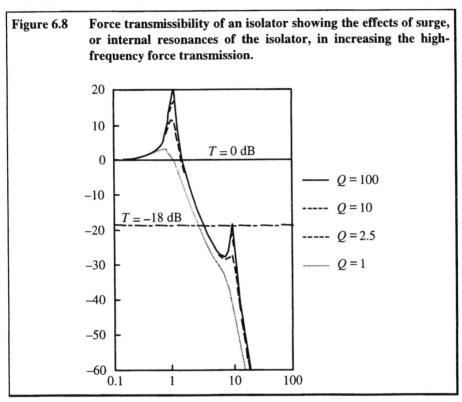

6.4
EFFECT OF FOUNDATION MOBILITY ON TRANSMITTED FORCE

The graphs in Figures 6.7 and 6.8 are computed assuming that the base under the isolators is rigid, which is of course an idealization. The transmitted force under the condition that there is no isolator and the base is rigid is the blocked force L_{bl}, which is the force transmitted to the foundation; the "hard mount" force shown in the mobility diagram in Figure 6.9a is therefore

$$L_{hd-mt} = L_{bl}Y_{int}/(Y_{int} + Y_{found}) \equiv L_{bl}Y_{int}/Y_{junct}$$

where Y_{junct} is the total junction mobility. If we now place an isolator of mobility Y_{isol} (we assume that the mass of the isolation system can be ignored), then the flow of force L_{isol} through the isolator is directly applied to the foundation, and using the mobility diagram in Figure 6.9b, we get $L_{isol} = L_{bl}Y_{int}/(Y_{junct} + Y_{isol})$. The junction mobility is simply increased by the isolator mobility. To see the effect of the isolator on the transmitted force, we take the ratio

$$\frac{L_{isol}}{L_{hd\ mt}} = \frac{Y_{junct}}{Y_{junct} + Y_{isol}} \tag{6.3}$$

which indicates to us the perhaps intuitive notion that the isolator will not provide a reduction in transmitted force unless the mobility of the isolator is large compared to the mobility already present at that junction.

The result in Equation 6.3, so easily obtained, is of great importance in thinking about the effectiveness of changing the properties of a junction to reduce noise transmission. Adding an isolator (which might take the form of a resilient gasket in a product application) will not be effective if either of the connected components has high mobility. That is why it is difficult for isolating components to have much effect in reducing the vibration of lightweight covers. We shall see a similar result to that of Figure 6.3 when we discuss transmission between connected structures in Chapter 7.

Figure 6.9   **Mobility diagrams used to determine the effect of foundation and internal (machine) mobility on the performance of an isolation system. The mobility of the isolator must be large compared to the junction mobility to be effective.**

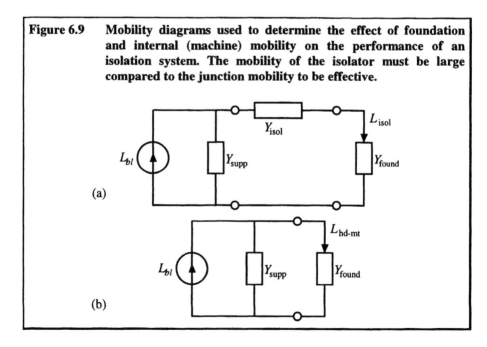

6.5
BLOCKING TRANSMISSION AT HIGH MOBILITY JUNCTIONS; BLOCKING MASSES

When the junction mobility is large, perhaps because the system is made of lightweight components, then a blocking mass may be used. The diagram in Figure 6.10 is relevant to this case. From the diagram, we can see that in order to drain off enough force to reduce the transmission to the foundation, the mobility must satisfy $Y_m \ll Y_{int}$, Y_{found}. In this case, analysis of the diagram indicates that

$$L_{found} = L_{bl} Y_m / Y_{found}$$

so that the ratio of mass to foundation mobility directly determines the reduction in transmitted force.

Unfortunately, in some instances substantial mass at the junction may be required to achieve a desired reduction in transmission. One way to accomplish this without penalty is to configure the product so that mass that is already present can be moved to achieve the goal. In one example, a cooling fan was isolated from the thin flexible walls of a computer by using the heat sinks in the power transistors for the computer as added mass, at the same time placing the cooling fins for the heat sinks in the fan's airstream. Clearly such a possibility must be regarded as a "target of opportunity," as it takes a happy combination of circumstances to achieve such a benefit.

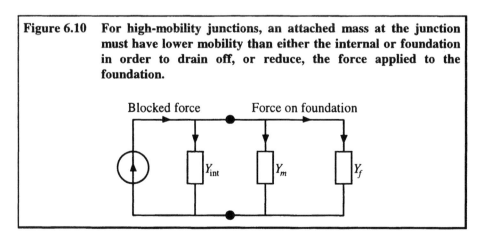

Figure 6.10   **For high-mobility junctions, an attached mass at the junction must have lower mobility than either the internal or foundation in order to drain off, or reduce, the force applied to the foundation.**

6.6
STRUCTURAL RESONANCES AS RESONANT MODES

If the single-dof resonator only applied to the situations discussed above, it would still have value, but the fact that it is also a model for each of the many resonances that structures and rooms have makes it an extremely powerful concept. We shall explore the structural aspects of this in more detail in Chapter 7, but here we will discuss why that should be so.

The resonator was described in Figure 6.1 as a combination of a mass, a spring, and a damping element. But already in Figures 6.3 and 6.4 that concept was broadened to identify mass elements with the storage of kinetic energy, stiffness elements with the storage of potential energy, and damping elements with the dissipation of energy. In that example, the kinetic energy KE of the machine was distributed throughout the machine and the cabinet, while the potential energy PE was distributed throughout the cabinet alone. The dissipation of energy Π_{diss} also occurred throughout the cabinet structure.

We can define element values once we define a reference velocity, and the values of the elements will depend on the velocity we choose because it is the energy of vibration that is fixed. In the example of Figure 6.3, the velocity chosen was the vibrational velocity of the sewing machine center of mass. For the resonant modes of a structure or sound field, we might instead use some space-time rms velocity and displacement. In any case, we define the mass as $m \equiv 2\text{KE}/\langle v^2 \rangle$, the stiffness as $K \equiv 2\text{PE}/\langle y^2 \rangle$, and the resistance as

$$R \equiv \Pi_{\text{diss}}/\langle v^2 \rangle.$$

These parameters obviously have values that depend on the reference velocity (or acceleration or displacement) and are equally applicable to either the resonance of a single dof resonator with lumped parameters or the resonant modes of a structure or room. Then, once we have defined the velocity variable, the kinetic, potential, and dissipated energies define the "element parameter" values.

CHAPTER 7

Multi-DOF Response; Transmission Between Structures

This chapter attempts to deal with the meat of the response problem in practical structures. We begin by relating the response of resonant modes to the single-dof system of the preceding chapter. Then the kinds of vibrational waves of potential importance are discussed. Next, different kinds of analyses are discussed; FEA, transfer matrix, and SEA. Applications of the methods to path analysis, vibration-induced fatigue, and source ranking complete the chapter.

7.1
MULTIMODAL RESPONSE, AVERAGE MOBILITY

All structures, regardless of the materials from which they are made, or how they are made up from simpler components such as plates or beams have resonant modes of vibration. It is rare that the structures have resonances that can be calculated from simple formulas, but their resonance frequencies and the corresponding vibrational (mode) shapes can be measured or computed using widely available hardware and software.

The hardware approach uses a vibration exciter, such as an impact hammer or electrodynamic shaker, and a motion sensor, typically an accelerometer or laser velocitimeter, to measure the induced vibrations, and a frequency analyzer to compute the transfer function between the source and reception locations. Transfer functions are measured for a variety of source and observation locations, and these are assembled into a software program that calculates mode shapes and resonance frequencies.

The computational approach starts with geometric and materials information about the structure and creates a "mesh model" of the structure. This may be done painstakingly from drawings or more easily by using software that converts CAD (computer-aided design) drawings directly into 3-D mesh models. The material properties for the components are specified (density, elastic moduli) and the FEA (finite element analysis) software then computes the natural modes and their resonance frequencies. The FEA program does not usually compute the damping loss factor for the modes. That information is supplied separately by the user.

Each mode of a structure therefore has a mode shape $\psi_m(x)$ and resonance frequency $\omega_m = 2\pi f_m$ where ω_m is measured in radians/sec (radian frequency) and f_m is measured in cycles/sec or hertz (Hz). From the discussion at the end of Chapter 6 we recall that the modal mass depends on the way we define the velocity of vibration. If it is the space-time ms, the modal mass equals the structural mass M. By definition, then, the modal stiffness is $K_m = \omega_m^2 M$ and the modal resistance is $R_m = \omega_m \eta_m M$, where η_m is the modal loss factor, all quantities defined in a manner completely analogous to the single-dof resonator. The parameters M, ω, and η are preferred because they can be directly related to measurements of mass, resonance frequencies, and damping.

When a concentrated (localized or "point") force is applied to the structure, then the effective force on a mode is proportional to the amplitude of that mode at that location x_s. Similarly, the vibration amplitude at the point of observation x_0 is proportional to the modal amplitude at that location. If we calculate the response at location x_0 due to all modes, we get the general equation for a transfer function (transfer mobility in this case)

$$Y_{0s} = \frac{j\omega}{M} \sum_m \frac{\psi_m(x_s)\psi_m(x_0)}{\omega_m^2 - \omega^2 + j\omega_m\omega\eta} \tag{7.1}$$

which, except for the mode shape weighting factors, is the same as Equation 6.1. But now the kinetic and potential energies and the dissipation of energy are all distributed over the structure instead of being concentrated in lumped elements.

The modal analysis software referred to above takes experimental data for Y_{0s} and attempts to fit it to a function like that in Equation (7.1). When the resonance frequencies are well separated, then this task is much easier that when modal responses overlap in frequency. That occurs when the average frequency separation between modes $\langle \delta \omega \rangle$ is comparable to or smaller than the modal bandwidth $\pi \omega \eta / 2$. The real part of the drive point mobility Y_{ss} (found by setting $x_0 = x_s$) is called the input conductance and, as in Chapter 6, it determines the power absorbed from the source. This function is sketched in Figure 7.1.

The graph in Figure 7.1 shows an important point, that power input to a structure occurs only a modal resonance frequencies. Therefore, the greater the number of modes, the greater the power injected into the structure. Because the "area" of each conductance peak is its maximum value, $\psi_m^2 / \omega_m \eta_m M$ times the width $\pi \omega_m \eta_m / 2$ distributed over the spacing $\langle \delta \omega \rangle$ gives an average conductance

$$\langle G \rangle = \pi \langle \psi^2 \rangle / 2M \langle \delta \omega \rangle.$$

This shows that the conductance and input power is greatest where the modal amplitudes are large, so that components that create force on the structure should be located where modal amplitudes are small to minimize vibratory energy.

7.2
STRUCTURAL RESPONSE IN TERMS OF WAVES

Modes of vibration are a very general and powerful descriptions of structural response, but the basic mechanics of the response are not very apparent. A wave description of the response gives a better idea of the mechanics. The three types of waves in structures of interest to us are longitudinal, shear, and flexural (bending). The connection between modes and waves is primarily through the wave speed, because as we shall see, the average spacing is inversely related to speed, so a lower wave speed means more modes and greater structural conductance.

Also, the impedance of the wave is $\propto \rho c$ where ρ is the structural density and c is the wave speed. Because all waves travel through the same structure, the speed alone determines relative wave impedance for different wave types. High-speed waves have high impedance, low-speed waves have lower impedance. High-impedance waves travel easily through high-impedance junctions and are reflected by low-impedance junctions, vice versa for low-impedance waves.

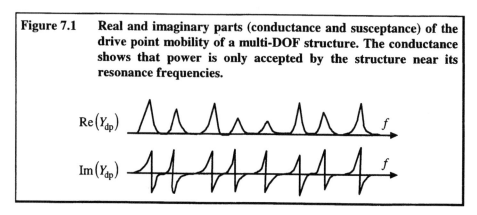

Figure 7.1   **Real and imaginary parts (conductance and susceptance) of the drive point mobility of a multi-DOF structure. The conductance shows that power is only accepted by the structure near its resonance frequencies.**

Figure 7.2 is a sketch of the vibrations for a longitudinal wave, which in steel or aluminum has a speed of 5100 m/sec. This wave is in-plane (vibrations are parallel to the free surfaces) and longitudinal (vibrations are parallel to the direction of propagation). Because of its high speed, this is a high-impedance wave and travels through the frames of a machine with little reduction, but as we shall see, its modes are relatively sparse.

Figure 7.3 is a sketch of the vibrations for a shear wave, which has a speed of about 3000 m/sec for steel or aluminum. This wave is also in-plane because its vibrations are parallel to the free surfaces. It is transverse because the vibrations are perpendicular to the direction of propagation. In a beam or rod, this wave corresponds to a torsional wave. Because both waves are in-plane, they are not strongly excited by forces perpendicular to the surfaces (such as sound wave pressures). In addition, these vibrations displace the air only slightly, so in-plane vibrations do not radiate much sound.

Figure 7.2   **A longitudinal wave has vibrations parallel to both the direction of wave travel and the free surfaces of the panel or beam. This is the fastest wave in the structure.**

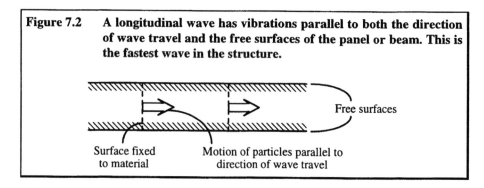

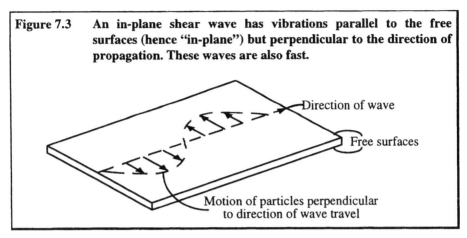

Figure 7.3   An in-plane shear wave has vibrations parallel to the free surfaces (hence "in-plane") but perpendicular to the direction of propagation. These waves are also fast.

Direction of wave

Free surfaces

Motion of particles perpendicular to direction of wave travel

Figure 7.4 is a sketch of a bending or flexural wave. This wave has out-of-plane vibrations, and its vibrations are transverse, perpendicular to the direction of propagation. These waves are fairly slow; the formula for steel or aluminum is $c_b = 3\sqrt{f(\text{Hz})h(\text{mm})}$ m/sec, where h is the plate thickness in mm. For example, if $h = 4$ mm and $f = 400$ Hz, the bending wave speed is 120 m/sec. These waves are very slow compared to longitudinal and shear waves. Their low impedance means they are reflected at junctions, and their low speed means that most of the resonant modes will be flexural. Because these vibrations are perpendicular to the free surface, these waves are readily excited by transverse forces such as sound wave impingement or mechanical impact.

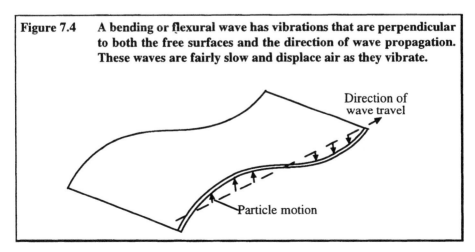

Figure 7.4   A bending or flexural wave has vibrations that are perpendicular to both the free surfaces and the direction of wave propagation. These waves are fairly slow and displace air as they vibrate.

Direction of wave travel

Particle motion

7.3
DETERMINISTIC MODELS FOR STRUCTURAL RESPONSE—CANONICAL STRUCTURES AND FEA

There are two types of deterministic models used in product design and analysis. The first is the set of models that have analytical solutions, such as simply supported beams and plates and rectangular rooms. These elements do not really exist in practice, but some of the analytical results can give good guidance to the behavior of actual structures. We call these "canonical" elements, because their solutions provide basic formulas that reveal fundamental properties of structures, even if actual structures do not have such solutions.

Finite element analysis (FEA) is analytical in the sense that the computational coding is based on analytical representations of the mesh elements. Because nearly any structure can be divided up into these elements, FEA is widely applicable. However, analytical formulas for frequencies or mode shapes do not result from the analyses, so that any modification in dimensions or material properties requires a totally new calculation. As there are natural variations that occur in manufacture, such repeated calculations may be desired. FEA models for products can be very elaborate, so the cost of repeated calculations may be prohibitive.

A simple but useful example of a canonical calculation is the flexural modal density of the rectangular simply supported plate shown in Figure 7.5, for which analytical functions for mode shapes and resonance frequencies are available. The mode shapes are

$$\psi_M(x) = 2\sin k_1 x_1 \sin k_2 x_2 \tag{7.2}$$

where $k_i = m_i \pi x_i / L_i$, $i = 1.2$; and the resonance frequencies are

$$\omega_M = \kappa c_l (k_1^2 + k_2^2) = \kappa c_l k^2 \tag{7.3}$$

where κ is the radius of gyration for the plate cross section and c_l is the longitudinal wave speed. The modes of this plate can be represented by the grid points shown in Figure 7.6, and the number of modes that resonate between k and $k + \Delta k$ is

$$N = \frac{\pi\, k\Delta k}{2} \bigg/ \frac{\pi^2}{L_1 L_2} = \frac{k\Delta k A_p}{2\pi} = \frac{\Delta\omega\, A_p}{4\pi\kappa c_l} \tag{7.4}$$

and the average modal separation, introduced above, is

$$\langle\delta\omega\rangle = \frac{\Delta\omega}{N} = \frac{4\pi\,\kappa c_l}{A_p} \tag{7.5}$$

If we introduce this modal separation into the expression above for average conductance, we get

$$\langle G\rangle = \frac{\langle\psi^2\rangle}{8\rho_s \kappa\, c_l} \tag{7.6}$$

where the average structural density is $\rho_s = M/A_p$. Equations 7.5 and 7.6 are examples of relationships derived from a idealized canonical model that are nevertheless useful for plate structures in general. These relationships, in terms of parameters that do not depend on the plates being rectangular or simply supported, have been found to be quite accurate for plates of various shapes and boundary conditions.

Figure 7.5 **A rectangular simply supported plate. Although an idealization, the results of some calculations based on this model can be applied to realistic panel structures.**

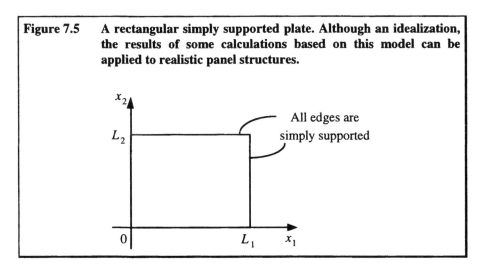

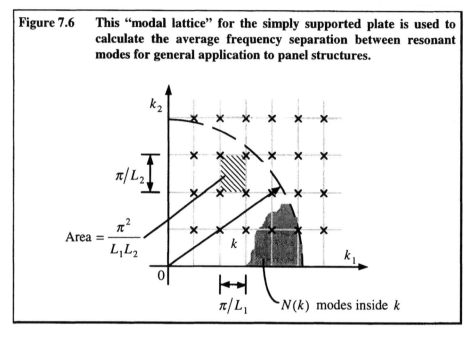

Figure 7.6 **This "modal lattice" for the simply supported plate is used to calculate the average frequency separation between resonant modes for general application to panel structures.**

As indicated above, there are few restrictions on the complexity of shapes that can be dealt with using FEA. Figure 7.7 shows an experimental sewing machine structure. Once the mode shapes and resonance frequencies are determined using FEA, equations like Equation 7.1 can be used to compute structural response to mechanism forces. Because of the commercial importance of this method, books, seminars, and online support are available for FEA procedures.

7.4
TRANSFER FUNCTION METHODS; EXPERIMENTAL AND ANALYTICAL

If the transfer functions for individual structural components can be determined, either from an analytical expression like Equation 7.1 or from experimental data, then the transmission of vibration when the structures are joined together can be calculated. Suppose we know the transfer functions for structure "a", Y_{12}^a from a set of inputs '1' to outputs '2' so that the force vector F_1 produces free velocities $V_2^{free} = Y_{12}^a F_1$. If a structure "b" is now connected to "a" at '2' (locations '3' on "b"), then the interaction force F_3 results. This force also produces local velocities at the connection given by

$$V_3 = Y_{33}^b F_3 = V_2^{\text{free}} - Y_{22}^a F_3 = Y_{12}^a F_1 - Y_{22}^a F_3, \tag{7.7}$$

or $F_3 = (Y_{22}^a + Y_{33}^b)^{-1} Y_{12}^a F_1$. Finally, using $V_4^{\text{free}} = Y_{34}^b F_3$, we get the transfer mobility for the combined system a + b as

$$V_4^{\text{free}} = Y_{34}^b (Y_{22}^a + Y_{33}^b)^{-1} Y_{12}^a F_1 \equiv Y_{14}^{a+b} F_1 \tag{7.8}$$

The mobility TF's that enter Equation (7.8) may be either the result of a measurement, an analysis, or both. If there is a (nearly) massless joining element for isolation Y_{isol}, then its mobility can be directly added to either Y_{22}^a or Y_{33}^b without changing the argument to produce a slightly modified formula for the overall TF;

$$Y_{14}^{a+b} = Y_{34}^b (Y_{22}^a + Y_{33}^b + Y_{\text{isol}})^{-1} Y_{12}^a \tag{7.9}$$

again showing that the transmission of vibrations will not be strongly affected by using an isolator unless the isolator mobility Y_{isol} is large compared to the mobility $Y_{22}^a + Y_{33}^b$ already present at the junction (the total junction mobility). When there are multiple dofs at the junction, then these statements must be interpreted in terms of matrix equivalents.

This way of combining the dynamics of structural components is not convenient when we wish to consider more than two structural components connected in tandem. When structures are arranged in a chainlike assembly, then a "T-matrix" method works well. Because the T-matrix elements are related to the transfer mobilities, the same experimental or analytical methods continue to apply. A typical T-matrix system component[12] is

$$\begin{pmatrix} V_n \\ F_n \end{pmatrix} = T_n \begin{pmatrix} V_{n-1} \\ F_{n-1} \end{pmatrix} = T_n T_{n-1} \begin{pmatrix} V_{n-2} \\ F_{n-2} \end{pmatrix} \cdots \tag{7.10}$$

and

$$T = Y_{12}^{-1} \begin{bmatrix} Y_{11} & Y_{12}^2 - Y_{11} Y_{22} \\ 1 & -Y_{22} \end{bmatrix}. \tag{7.11}$$

By using the T-matrix, a complete string of structures can be combined by rapid matrix manipulation. Then, if the system transfer mobility is desired, reference to Equation 7.11 shows that that function will be the inverse of the element in the lower left corner of the matrix, and drive point mobilities will be determined by the diagonal components of the matrix.

Figure 7.7   A finite element structural analysis of an experimental sewing machine structure shows the vibrational shape of one of its low-frequency modes that might be excited by mechanism imbalance.

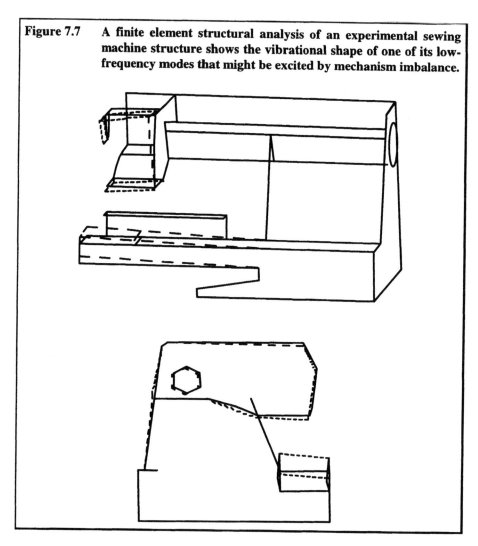

7.5
APPLICATION OF THE TF METHOD TO VIBRATION PATH ANALYSIS

Prescribed displacement inputs at the rear tires of a vehicle produce vibrations in the rear suspension that travel into the vehicle through the suspension mounts that terminate near the rear shelf of the interior and on through the trailing links in the suspension to the driver's ear, as shown in Figure 7.8. An experimental determination of the transmission through the trailing link path and the overall TF that includes both paths was carried out to determine which path was controlling the transmission.

The overall TF from the tire tread to the driver's ear was measured and is graphed in Figure 7.9a. The TF is expressed as the ratio of p_6 to a_1. Because the connection between the suspension and the vehicle could not be removed, it was necessary to determine the TF through the trailing link by examining its components, as shown in Figure 7.10, and using Equation 7.9 to determine the path TF. This path is divided into two sections, from the tire tread '1' to the link end at the car body '4,' and from the car body connection '5' to the driver's ear '6'. The overall tread to ear TF through the trailing link is therefore

$$\left. \frac{p_6}{a_1} \right]_{\text{trailing-link}} = \frac{(a_4/f_1)(p_6/a_5)}{(a_4/f_4 + a_5/f_5)} \qquad (7.12)$$

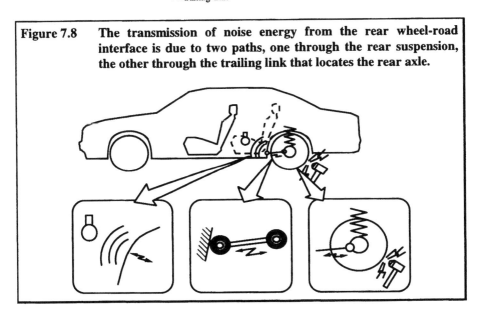

Figure 7.8 **The transmission of noise energy from the rear wheel-road interface is due to two paths, one through the rear suspension, the other through the trailing link that locates the rear axle.**

Figure 7.9   **A comparison of**

(a) the total transfer function from force at the tire surface to sound pressure at the driver's ear with

(b) that due to the trailing link only shows the frequencies where each of the two paths contribute.

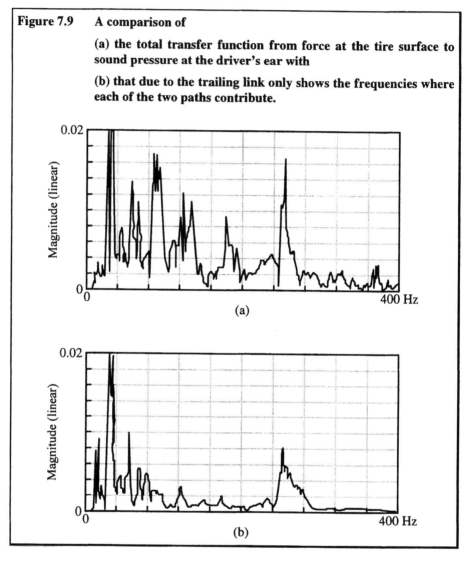

(a)

(b)

The trailing link TF calculated in this manner is shown in Figure 7.9b, and it shows that the noise in the frequency range near 35 Hz and 250 Hz can be accounted for by the training link path, but the noise at frequencies near 60, 90, 130, and 180 Hz are carried by the suspension path.

Figure 7.10   **Diagram showing how the measurements on the trailing link path transfer function is determined.**

(a) The part of the path from the tire surface to the chassis connection.

(b) The part of the path from the tire surface to the chassis connection.

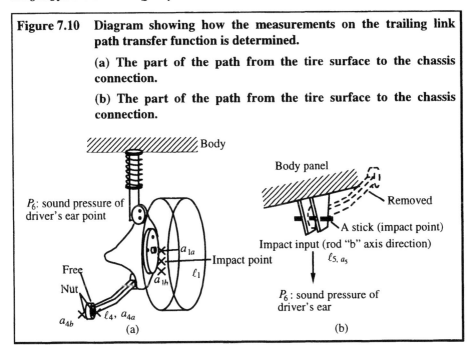

7.6
NONDETERMINISTIC METHODS; SEA, POWER FLOW, OTHER

Actual products when manufactured show a surprising amount of variability in their vibrational response and sound radiation. Products like sewing machines that produce sound over a wide frequency range may have a standard deviation (sd) in their overall sound output of 1.5–2 dB, with individual tones in their spectra varying by as much as 10–20 dB. Sample data for the A-weighted sound from a sample of 48 sewing machines is shown in Figure 7.11. In light of this variability, one may well wonder just what a prediction of sound or a measurement of a prototype means in terms of predicting the sound from the final product.

In addition to the design challenge, variability also has production implications, because if a certain sound level is specified for the product, then the *average* sound produced by the product will have to be 3–4 sd's below the pass-fail limit to have an acceptable failure rate. Variability is also a customer satisfaction issue, as more variability increases the chance of a customer complaining about a noisy product, and product repairs or replacements in the field are expensive.

Figure 7.11   **This histogram of the sound levels radiated by a group of sewing machines on the production line shows that their sound varies with a standard deviation of about 1.5 dB.**

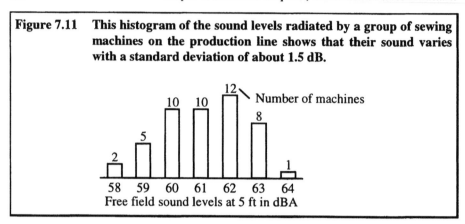

Variability is one aspect of product behavior, complexity is another. The TF shown in Figure 7.12 is the transfer *accelerance* for a diesel engine structure, the ratio of acceleration at the observation location to applied force at the source location. The outstanding feature of this data is its complexity; the number of resonant modes is large, and the range of amplitudes between the peaks and notches in the TF is 20–30 dB.

Figure 7.12   **This transfer accelerance data for a diesel engine casting shows the complexity and large dynamic range of its vibration TF values.**

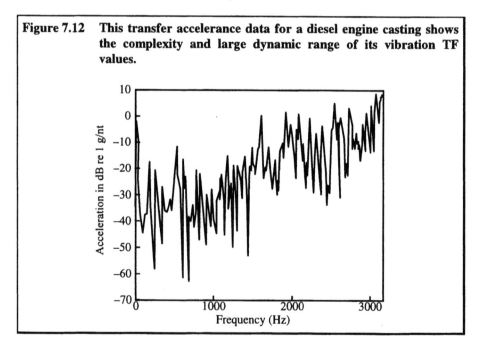

If this TF is measured for a group of engines, and the mean and sd are computed, then the results are as shown in Figure 7.13. The average value of TF is somewhat smoother than the single TF in Figure 7.12. The sd at low frequencies is a few dB, but this increases to 7–8 dB as frequency is increased. The variability may arise from within the engines themselves or from the way they are mounted, but the variability in response is present. This large an sd means that 5 percent of the samples will have TF's that differ by more than 30 dB at any frequency. A glance at Figure 7.12 shows that small frequency shifts in resonance frequencies could easily produce such variations.

Statistical energy analysis (SEA) was developed to deal with situations like that in Figure 7.13.[13] It takes the view that one is estimating the behavior of a population of structures or sound fields, not a single sample. It also uses a result from random vibration theory that is reminiscent of heat flow; the energy flow from one resonance in the system to another resonance is proportional to the difference in their energies (similar to temperature). The parameter governing this flow is a *coupling loss factor*. If we consider two connected structures, 'a' and 'b', and if 'a' is excited by noise in a band $\Delta\omega$, but 'b' is only excited due to its connection to 'a', then according to SEA, the power flow in terms of their energies is

$$\Pi_{ab} = \omega\eta_{ab}E_a - \omega\eta_{ba}E_b = \omega\eta_b E_b \qquad (7.13)$$

Figure 7.13 **Mean and standard deviation statistics for engine vibration TF's show large variability of response among "identical" engines.**

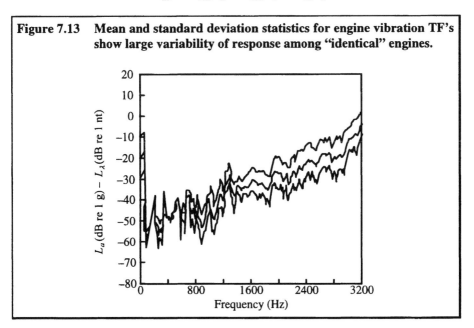

where ω is the center frequency of the band and η_{ab} is the coupling loss factor from 'a' to 'b.' Because 'b' is not excited by an external source, the transmitted power must be dissipated in 'b.' The two coupling loss factors are not independent, but are related by the mode counts in the structures; $\eta_{ab}N_a = \eta_{ba}N_b$.

The coupling loss factors are found in a variety of ways: junction mobilities, wave transmissibility at junctions, and experimental testing. As indicated above, if the vibrations are damping controlled (resonant), then

$$E = M\langle v^2\rangle_{x,\,t}$$

and one can then solve Equation 7.13 for the ratio of velocities (or accelerations)

$$\frac{\langle v_b^2\rangle}{\langle v_a^2\rangle} = \frac{M_a\eta_{ab}}{M_b(\eta_{ba}+\eta_b)}. \tag{7.14}$$

Using relations like this, we can estimate the ms response throughout a structure using simple algebraic formulas that include the SEA parameters. Software packages are now available that allow one to carry out SEA-based predictions of response in systems of varying size and complexity.

In the preceding, we were not explicit about whether the systems in question are structures or sound fields or whether the description is based on wave or resonant mode descriptions. A description particular to structures and based on wave concepts is the "power flow method."[14] Figure 7.14 shows plate 'a' excited by a noise source that produces a reverberant wave field in the plate. These waves produce a ms velocity on the plate $\langle v_a^2\rangle$, and an energy of vibration $E_a = M_a\langle v_a^2\rangle$. This energy strikes the boundaries of the plate at as rate $v_a = c_a/d_a$, where c_a is the energy or group speed of the waves and $d_a = \pi A_a/L_{\text{junct}}$ is the mean free path (average distance) for energy packets in the plate to reach the junction of length L_{junct} with plate 'b'. Therefore, the power incident on the boundary with plate 'b' is $\Pi_{\text{inc}} = v_a E_a$. The fraction of this power that is transmitted through the boundary is the *transmissibility* τ_{ab}. But vibrational energy in plate 'b' will flow back into plate 'a' through the junction, so the net power flow from 'a' to 'b' will be

$$\Pi_{ab} = v_a E_a \tau_{ab} - v_b E_b \tau_{ba}. \tag{7.15}$$

The similarity with Equation 7.13 is obvious. In the case of plates, a correspondence can be drawn between the SEA quantities modal energy and coupling loss factors and the power flow quantities incident power and transmissibility: $E_a/N_a = \Pi_a^{inc}/N_a v_a$ and $\eta_{ab} = (v_a/\omega)\tau_{ab}$.

Another way of determining the statistical behavior of an assemblage of systems is to carry out repeated calculations on a deterministic system that has parameters that can be readily varied. This sometimes goes under the title of a *Monte Carlo* method. As a deterministic calculation of vibration transmission or sound radiation for a product may be quite lengthy, one must be judicious in attempting such calculations. Variations in response or radiation are not like dimensional tolerance stackup, where one can easily calculate a worst case by using the extreme dimensions of assembly components.

An illustrative but idealized example of this procedure is carried out on the structure shown in Figure 7.15. In this case, the two simply supported plates have mode shapes and resonance frequencies that are known and do not have to be calculated using computational methods (see Equations 7.2 and 7.3), but if they were not known, a FEA procedure could be used to produce these quantities. The model also allows calculations of sound radiation, but we will defer that discussion to Chapter 8.

The two plates are connected by a circle of five joining elements, which we might think of as spot welds or rivets, each modeled as a spring and damper. These elements are assumed to have a distribution of parameter values and locations because of variations in the production process. The plates are complicated dynamically, with many modes of vibration, and are assumed deterministic and remain unchanged for every realization. The joining elements are simple dynamically, but are assumed to be randomly located and to have random parameter values.

Figure 7.14 **The energy transfer between coupled plates is represented by a transmissibility, a concept analogous to sound transmission through a wall between two rooms.**

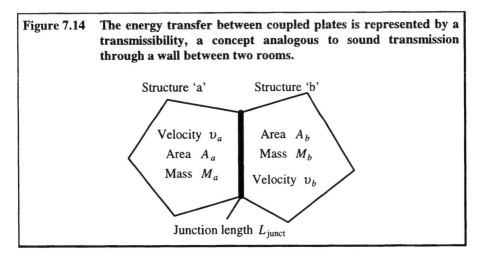

Structure 'a'   Structure 'b'

Velocity v_a Area A_b
Area A_a Mass M_b
Mass M_a Velocity v_b

Junction length L_{junct}

Figure 7.15   **This sketch of two plates connected by a ring of rivets is the model for a system of deterministic structures coupled by randomly varying connections.**

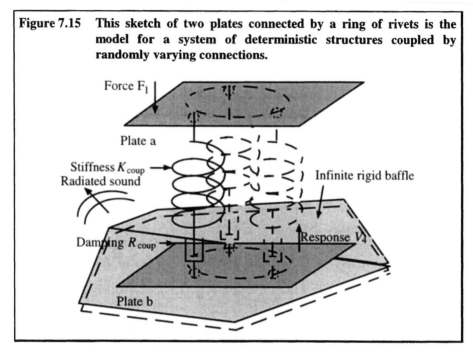

We use the result in Equation 7.9 to calculate the vibration of plate 'b' when a force is applied to plate 'a' at a particular location. The isolator mobility in this case becomes the mobility of the joining elements. The Monte Carlo calculations can proceed rapidly and efficiently because the mode shapes and resonance frequencies determine the transfer functions for the two plates. The representations of their transfer and drive point mobilities involving analytic mode shapes would be replaced in a less ideal situation by FEA derived mode shapes.

A sample outcome of calculations for an ensemble of 64 realizations of the system shown in Figure 7.15 is presented in Figure 7.16. We see that the response is fairly repeatable for the first few modes, but becomes more random and diffuse as frequency is increased and we get to the higher modes. This result is quite consistent with data on the transfer mobility in actual structures made on a production line, exemplified by the data in Figure 7.13.

Because we know the plate and connecting element properties in this example, it is also possible to calculate a SEA response of plate 'b' to a force on plate 'a.' The average modal spacing for the two plates, their average drive point conductance, and their damping are all known. We can use an impedance-based formula for the coupling loss factor to complete the list of needed functions and parameters for an SEA estimate:

$$\eta_{ab} = 2N_{\text{conn}} \langle \delta\omega_a \rangle G_a G_b / \pi \omega \langle |G_a + G_b + Y_{\text{conn}}|^2 \rangle \qquad (7.16)$$

where the energy flow through each of the connecting elements is assumed to be independent and therefore additive. The SEA prediction of acceleration of plate 'b' due to a force on plate 'a' is then

$$\langle |Y_{14}|^2 \rangle = \pi N_a \eta_{ab} / 2\omega \, \Delta\omega \, M_a M_b (\eta_a \eta_b + \eta_{ab} \eta_b + \eta_{ba} \eta_a) \qquad (7.17)$$

This function is also graphed in Figure 7.16. We see that it provides a useful estimate of the vibration of the plate 'b', but it does not of course predict the variations about a mean value for the response. Parenthetically, note that it is reciprocal in the properties of the two plates, as it should be.

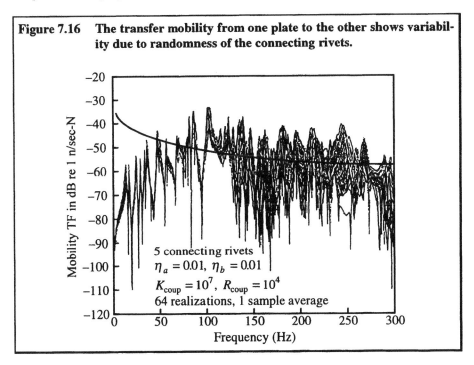

Figure 7.16   **The transfer mobility from one plate to the other shows variability due to randomness of the connecting rivets.**

Figure 7.17   A knitting machine uses needles driven by cams that cause abrupt accelerations and resulting vibrations leading to fatigue damage.

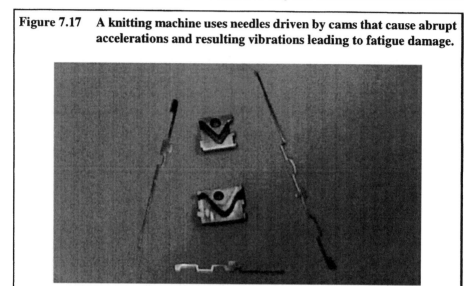

7.7
APPLICATION TO VIBRATION AND STRESS ANALYSIS; EXAMPLE OF A KNITTING MACHINE

Vibrations not only produce sound, they also produce stresses within product components that, if severe enough, can result in fatigue failure. Although fatigue damage is inherently a nonlinear process, such nonlinearities do not affect the dynamics for high-cycle fatigue, so linear theory can be used to estimate dynamic stresses, and then fatigue damage rate and life can be calculated using the appropriate model. In the following, we implicitly use *Miner's Rule*,[15] which says that damage accumulates in the material on a cycle-by-cycle basis.

An industrial knitting machine employs a large number of cam-driven needles in a circular arrangement. Such a needle is shown in Figure 7.17 along with a segment of the driving cam. The slot of the cam engages the protruding "butt" of the needle, causing the needle to undergo the path prescribed for the knitting pattern. The shape of the needle is intended to reduce the amplitude of any shock excitation from the cam getting to the free end where the hook is located.

Because the cam moves at a constant velocity, steps in slope of the slot produce steps in velocity and impulses of acceleration. Abrupt changes in curvature of the cam slot produce steps in acceleration and impulses of "jerk" of the needle. In either case, high-frequency excitation of the needle occurs, leading to needle vibrations that travel to the hook. Because the slender hook of the needle is near a free end of the needle, large bending vibrations occur at the hook, causing high stresses and fatigue failure.

Figure 7.18 shows the pattern of stresses for one of the lowest bending resonances of the needle at 10.8 kHz. The set of stress contours shows a substantial concentration of stress at the hook. The finite element analysis of the needle predicts from 10–15 needle resonances from 0–50 kHz. Not all of these resonances produce high stresses in the hook area, of course, but their responses can be combined to predict the time history of stress at the hook for a given rate of acceleration pulses from the cam.

A dynamic simulation of the excitation for a particular cam profile that has an impulse of jerk with a defined duration results in a time waveform for the stress in the hook area shown in Figure 7.19. The model for the material involved has an endurance limit, shown as a dashed line in Figure 7.19, which means that only stresses above that value contribute to damage. A histogram of stress amplitudes presented on a fatigue damage rate graph is shown in Figure 7.20. The rate of stress peaks above the endurance level allows one to compute damage rate, the inverse of which is fatigue life. But because of the very large number of stress cycles a knitting needle undergoes, the design goal must be to stay below the endurance limit for the material.

Figure 7.18 **Stress contours show concentration of stress near the area of the needle hook due to resonant mode response. These stresses may lead to fatigue damage and needle failure.**

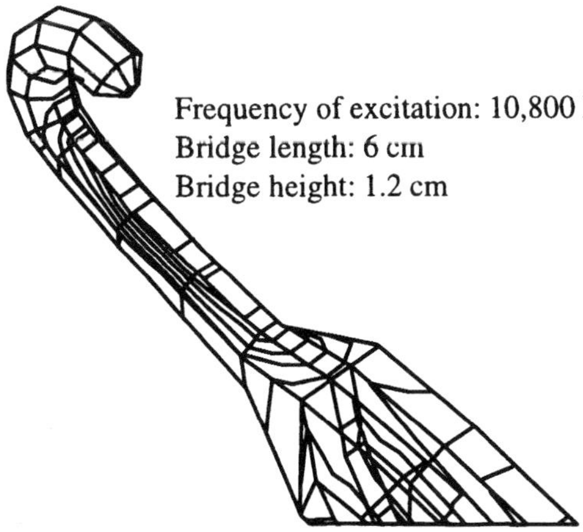

Frequency of excitation: 10,800 Hz
Bridge length: 6 cm
Bridge height: 1.2 cm

Figure 7.19   **Calculations of needle vibration due to cam profile lead to a synthesized hook stress time history, usable in calculating a fatigue damage rate.**

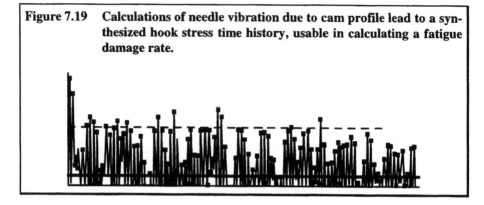

Figure 7.21 shows how the peak stresses depend on the rise time constant of the cam acceleration for three different needle designs. If one keeps the cam profile so that the rise time $\tau_r > 10^{-5}$ sec., then the peak stresses for these designs will stay below the endurance limit. This translates to a distance of $V_{cam}\tau_r$ along the cam face to change the curvature where V_{cam} is the cam velocity. If this distance is 1 mm, then the maximum velocity of the cam is $V_{cam}^{max} = 10^2$ m/sec. There is a speed (and therefore a production) limit based on how smooth the cam profile can be allowed to be.

Figure 7.20   **The S-N curve for the material used in the knitting needle and the stress peak histogram showing some levels above the endurance limit for the material indicates that there is accumulation of damage.**

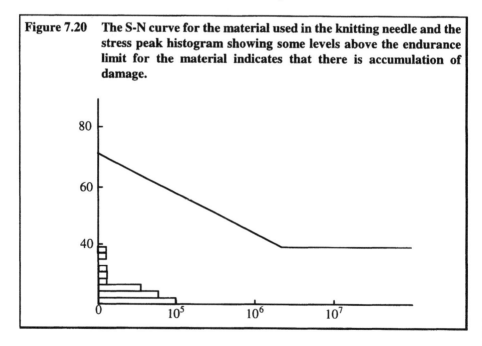

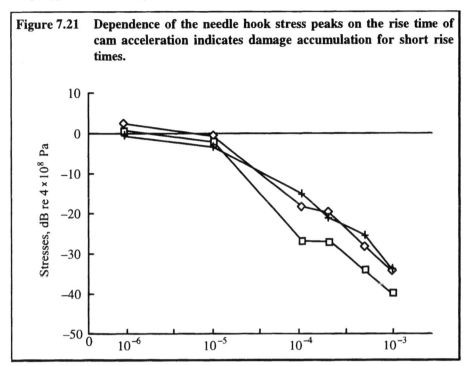

Figure 7.21 **Dependence of the needle hook stress peaks on the rise time of cam acceleration indicates damage accumulation for short rise times.**

7.8
RELATIVE CONTRIBUTION FROM SOURCES; DIESEL ENGINE EXAMPLE[16]

We discussed two sources of noise in a diesel engine in Chapter 4—combustion pressure and piston slap. We now see how these sources, along with the fuel injection system, can be combined to account for the total casing vibration of the engine. We will use measured data for source waveforms and transfer functions for each of the three sources, along with measured values of casing vibration to see how each contributes to the overall response.

Figure 7.22 shows the cylinder pressure waveform and its spectrum for a turbocharged six-cylinder 9.2-liter engine. This pressure produces force upward on the cylinder head and downward on the piston. A V-engine has a stiffer crankshaft than does an in-line engine, so the path of vibration from cylinder pressure to engine casing is weaker for a V-engine than for an in-line engine.

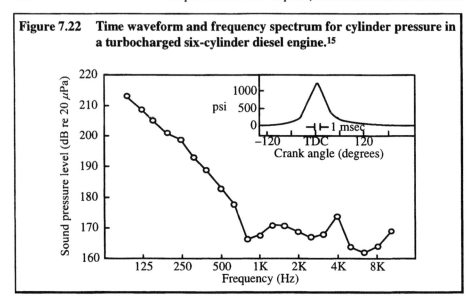

Figure 7.22   Time waveform and frequency spectrum for cylinder pressure in a turbocharged six-cylinder diesel engine.[15]

When the measured transfer functions for both paths are combined on an energy basis and applied to the pressure spectrum in Figure 7.22, we get the calculated casing vibration due to cylinder pressure shown in Figure 7.23. The figure also includes the measured casing vibration, and we see that we can only account for the vibration in the lowest frequency bands as due to combustion pressure.

The waveform and frequency spectrum of cylinder wall acceleration due to piston slap are shown in Figure 7.24. We note the two peaks in this spectrum at about 1 and 2 kHz, which are similar in value to the two peaks in the spectrum of piston slap that we discussed in Chapter 4, even though the engines are of quite different size. This indicates that as engines of different size are developed, their weight and stiffness values increase in proportion to each other.

When the transfer function relating cylinder wall to casing vibration is applied to the data in Figure 7.24, the estimated casing vibration due to piston slap alone is as presented in Figure 7.25. We see that piston slap appears to be the primary source of vibration in a frequency range from about 600 to 1500 Hz, with contribution also from 100 to 300 Hz. This relative dominance of piston slap over combustion pressure is probably not universal. Turbocharging tends to smooth out the combustion pressure somewhat, reducing the high-frequency content of the combustion pressure, and the V construction of the engine also reduces the contribution of combustion pressure to casing vibration.

Figure 7.23 Use of cylinder pressure to casing vibration TF allows an estimate of casing vibration due to combustion pressure. Comparison with measured engine vibration shows principal contribution to be at low frequency for the V-type engine.[15]

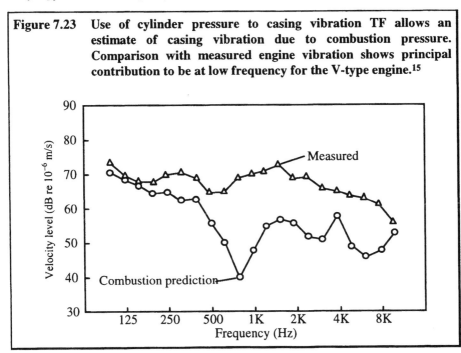

Figure 7.24 Cylinder bore acceleration waveform and frequency spectrum due to piston slap shows peaks at 1 kHz and 2 kHz due to piston-bore resonances.[15]

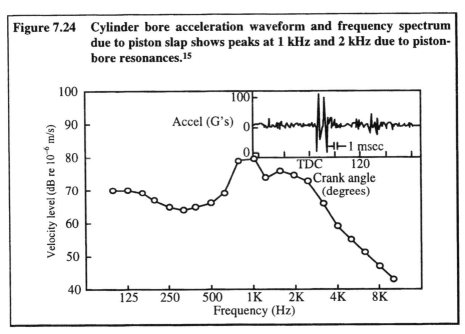

Figure 7.25   **Use of cylinder bore to casing TF shows that predicted piston slap vibration matches engine vibration at low and mid frequencies.[15]**

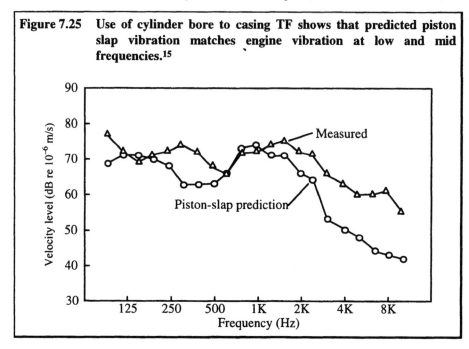

The motion of the cam follower in the fuel injection pump is presented in Figure 7.26. Both the velocity due to the design profile and actually measured velocity waveforms are presented. The spectrum of the measured velocity is also presented. It appears the cam and follower mechanism has a resonance at about 1 kHz. When the transfer function relating follower velocity to casing acceleration is applied to this data, an estimate for casing vibration due to injector action results, as presented in Figure 7.27. It seems clear that the fuel injector is responsible for the observed casing vibration above 2 kHz.

Finally, we can add the inferred vibration spectra for the three sources—combustion pressure, piston slap, and fuel injectors—and compare them with the measured spectrum of casing vibration. The result of this comparison is shown in Figure 7.28. Although the separate sources do not predict the overall measured response exactly, a comparison of this sort is normally considered to be excellent.

In any case, the example clearly illustrates the point that in many products, one cannot reduce the vibration or sound from a product without dealing with all sources. A particular source may be having an objectionable effect on sound quality, even if it is not a major contributor to overall noise level.

Figure 7.26 **Cam-plunger velocity waveform and its spectrum show high noise energy at and above 1 kHz.**[15]

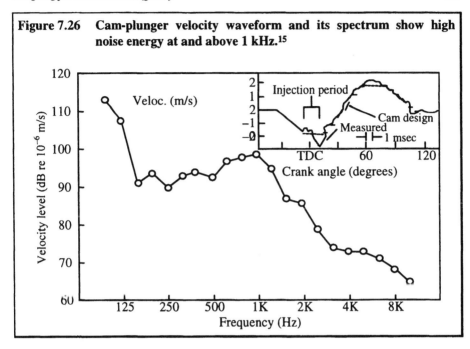

Figure 7.27 **Using the TF from injector plunger motion to casing vibration shows injector-induced vibration is important above 1 kHz.**[15]

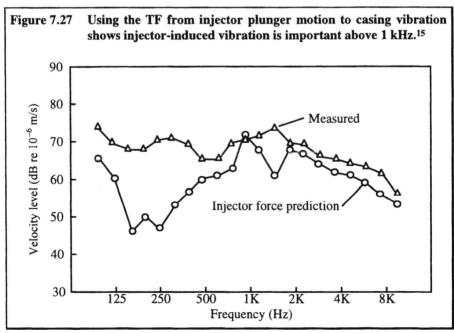

Figure 7.28   **Combining all sources provides a prediction of casing vibration that is remarkably close to the measured vibration during engine operation.[15]**

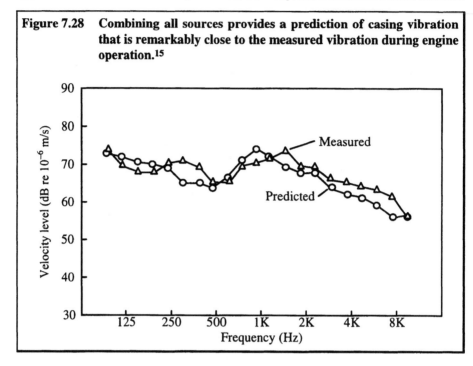

CHAPTER 8

Basics of
Sound Radiation

We have studied how motions and forces are generated by the various mechanisms in a product. In some cases, like the aerodynamic forces in Chapter 5, sound is generated directly by the forces, even if no structural vibration results. We have also discussed how the mechanisms produce vibrations in the structure, and these vibrations also radiate sound, quantified by a parameter called the radiation efficiency of the structure. Designing the structure for Sound Quality is, in part, designing the structure to have the "right" radiation efficiency.

8.1
ACOUSTICAL VARIABLES FOR PLANE WAVES OF SOUND, RADIATION EFFICIENCY

The simplest kind of sound wave in air is somewhat like the longitudinal stress wave discussed in Chapter 6. In the case of air, the bulk stiffness is γP_0 and the air density is

$$\rho_0 = 1.18 \ \text{kg}/\text{m}^3,$$

where $\gamma \equiv C_p/C_v \approx 1.4$ is the ratio of specific heats and $P_0 = 10^5 \text{ N/m}^2 (\equiv \text{Pa})$ is the atmospheric pressure. The speed of the wave is

$$c_0 = (\gamma P_0/\rho_0)^{1/2} \approx 345 \text{ m/sec}$$

at standard conditions. The (specific acoustic) impedance of the wave is $z_{\text{wave}} = \rho_0 c_0 = 407$ mks, which defines the ratio of pressure to air particle vibration velocity in the plane traveling wave.

If the wave is generated by a vibrating piston in a tube or pipe as shown in Figure 8.1, then the piston produces a velocity v in the air, accompanied by a pressure $p = \rho_0 c_0 v$, so the power absorbed by the wave is

$$\Pi_{\text{rad}} = \langle pv \rangle_t A_p = \rho_0 c_0 A_p \langle v^2 \rangle_t. \tag{8.1}$$

Although this formula deals with a very special case, we can use it as the basis for a much more general situation.

First however, let's note that Equation 8.1 does have some practical utility as it stands. In cases where we are willing to assume "plane-wave" like radiation (radiation from large stiff structures, for example), we can relate the velocity of vibration directly to sound intensity (power per unit area) or sound power. Figure 8.2 shows a third-octave spectrum of vibration velocity where the reference velocity is 1 m/sec on the left-hand scale and 5×10^{-8} m/sec on the right-hand scale. In air, the reference for sound pressure $p_{\text{ref}} = 2 \times 10^{-5}$ Pa and the intensity reference is $I_{\text{ref}} = 10^{-12} \text{ W/m}^2$. The result is that intensity and pressure levels are numerically equal for a plane propagating sound wave.

Figure 8.1   **A vibrating piston in a tube generates plane waves of sound in a medium of density ρ_0 and speed of sound c_0.**

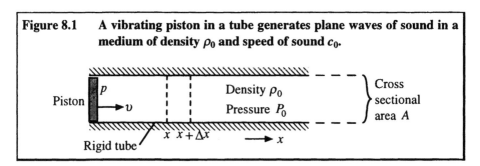

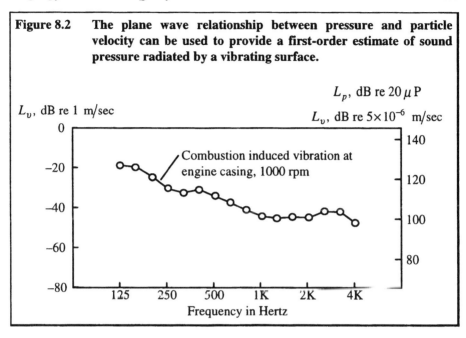

Figure 8.2 The plane wave relationship between pressure and particle velocity can be used to provide a first-order estimate of sound pressure radiated by a vibrating surface.

From Equation (8.1), the intensity is $I = \rho_0 c_0 < v^2 > = < p^2 > / \rho_0 c_0$ which relates velocity to intensity also. If we choose the velocity reference 5×10^{-8} m/sec, then the velocity, intensity, and pressure levels are all numerically equal, and the graph of velocity levels becomes an estimate for sound pressure and intensity levels as well. Noise control engineers trying to estimate the sound output of large machines at moderate to high frequencies find a formula like that in Figure 8.2 quite useful.

We can also use Equation 8.1 as the basis for a more general relation between vibration and sound. If we have an arbitrary vibrating structure as shown in Figure 8.3, then we can still define a space-time ms velocity

$$\left\langle v^2 \right\rangle_{x,t},$$

and an area for the structure A_S, so the radiated power can be expressed as

$$\Pi_{\text{rad}} = \rho_0 c_0 A_S \left\langle v^2 \right\rangle_{x,t} \sigma_{\text{rad}} \tag{8.2}$$

where the quantity σ_{rad} is a "fudge factor" to make the equation correct. This fudge factor is the radiation efficiency (or radiation factor) mentioned above, and it can be thought of as depending on two length scale comparisons:

$$\sigma_{rad} = \sigma_{geom}(D/\lambda_{sound}) \times \sigma_{dyn}(\lambda_{vib}/\lambda_{sound}) \qquad (8.3)$$

where the "geometric radiation efficiency" σ_{geom} depends on the ratio of the size of the radiator D to the wavelength of sound λ_{sound} and the "dynamic radiation efficiency" σ_{dyn} depends on the ratio of the wavelength of the vibration to the wavelength of sound. We discuss both of these factors in following sections.

8.2
SOUND IN PIPES; EXAMPLE OF A COMPRESSOR

We have shown how standing waves on a beam or in a plate result in resonances that act like resonators. A similar process exists for sound waves in piping systems. The two-stage air compressor shown in Figure 8.4 is connected to a pipe that dumps the air into a storage tank (accumulator). The source of air, the compressor, acts like a source of prescribed volume velocity and has a very high internal impedance. The opening of the pipe into the tank is a very small acoustic impedance, so the pipe has resonances when it is 1, 3, 5 quarter wavelengths long. These resonances can have an effect on the performance of the compressor.

Figure 8.3   **The radiation efficiency (factor) is an adjustment to the plane wave radiation law to allow sound power to be calculated or estimated for a vibrating structure of known area and ms velocity.**

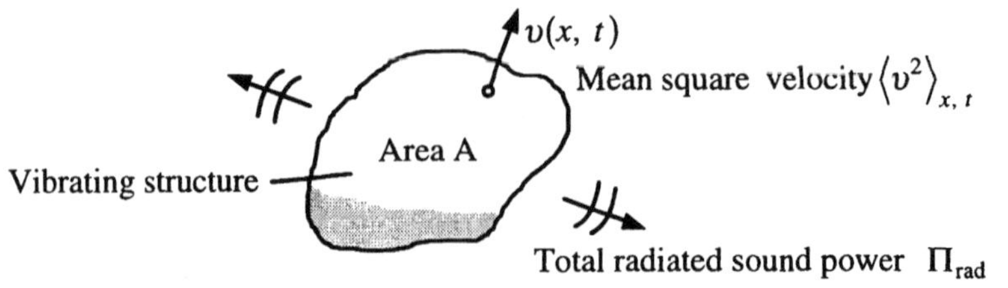

$v(x,t)$

Mean square velocity $\langle v^2 \rangle_{x,t}$

Area A

Vibrating structure

Total radiated sound power Π_{rad}

Figure 8.4   An industrial two-stage compressor dumps compressed air into a tank by a length of pipe that turns out to be critical.

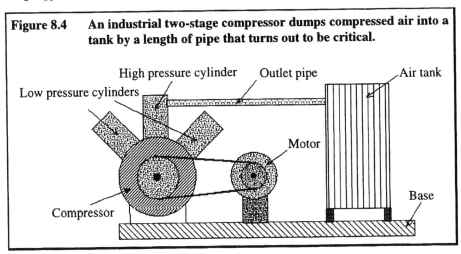

The diagram in Figure 8.5 shows the air pressure at the inlet manifold of the high-pressure stage, the high-pressure stage cylinder pressure, and the pressure in the outlet manifold. The inlet pressure to the high-pressure stage is fairly constant at about 30 psi. The exhaust valve in the cylinder head is designed to open when the cylinder pressure exceeds the exhaust manifold pressure by a few psi. In this case, the valve should open at about 105 psi, but because the pressure in the exhaust manifold is oscillating, the valve does not open and eject air into the manifold until the pressure reaches almost 120 psi (causing the "bump" in the exhaust manifold pressure), greatly reducing the efficiency of the compressor.

From the time scale at the bottom, we see that the period of the compressor is about 67 msec., corresponding to a frequency of 15 Hz, or 900 rpm. The oscillation in the exhaust manifold pressure is at a frequency of 30 Hz (with harmonics), so that every time the exhaust valve opens, the manifold pressure has risen, reducing the flow from the compressor. The piping resonance at 30 Hz corresponds to a wavelength of about 12 meters or a quarter wavelength of 3 meters, which is about the length of the pipe from the compressor to the tank.

Because the operating speed of the compressor is fixed at 900 rpm, the coincidence between the operating frequency and the piping resonance could be avoided by changing the length of the pipe. Another solution is to add a "Helmholtz resonator" to the piping at the end near the compressor as shown in Figure 8.6. This shifts the lowest resonance of the piping to a frequency of about 60 Hz (pipe length = half wavelength), which avoids the problem.

Figure 8.5   **The inlet pressure to the second stage of the compressor is fairly steady, but pressure oscillations in the outlet manifold oppose the airflow and reduce compressor efficiency.**

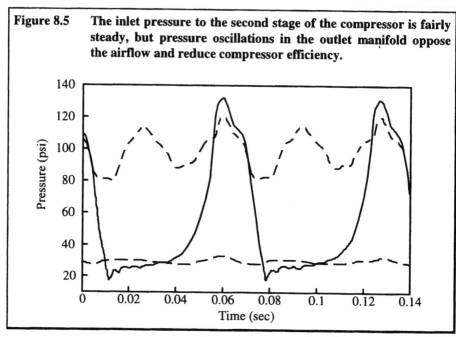

Figure 8.6   **The tendency of the connecting pipe to act as a one-fourth wavelength standing wave tube at 30 Hz causes the oscillations to have the right phase to oppose the exit flow from the compressor. A Helmholtz resonator connected at the manifold end of the pipe and tuned to 30 Hz eliminates the problem.**

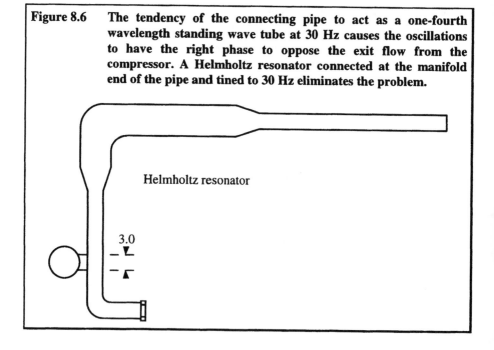

The fact that the piping resonance acts to make it more difficult for the compressor to inject air into the outlet manifold is an example of the acoustical resonance acting like a dynamic absorber. Dynamic absorbers are often used in machinery to restrict vibration at the machine's operating speed. When a mass-spring resonator like that shown in Figure 5.1 is set on a vibrating structure and driven by motion at its base, then at resonance its motion becomes very large, and it generates a force in the spring to oppose the vibration that is exciting it. In terms of the discussion in Chapter 6, the mobility of the resonator is very large when we drive the mass with a force, but the mobility is very small when we drive the resonator with motion at the base through the spring.

8.3
SOUND FROM A PULSATING SPHERE; THE GEOMETRIC RADIATION EFFICIENCY

The three-dimensional counterpart to the vibrating piston of Figure 8.1 in one-dimension is the pulsating sphere shown in Figure 8.7. This source is very similar to a plane wave source at high frequencies where the wavelength of sound is very small compared to the diameter of the sphere, so the radiation efficiency becomes nearly unit at high frequencies.

Although the radiation efficiency is unity, the sphere is still radiating a spherical wave, and the pressure has to be inversely proportional to the distance from the origin so that the intensity is inversely proportional to the square of the radius. In such a case, the product of intensity and area, which is the sound power, is independent of the radius at which it is calculated, which of course it should be.

Figure 8.7   **A uniformly pulsating spherical source is the model for a "simple," or monopole, source of sound.**

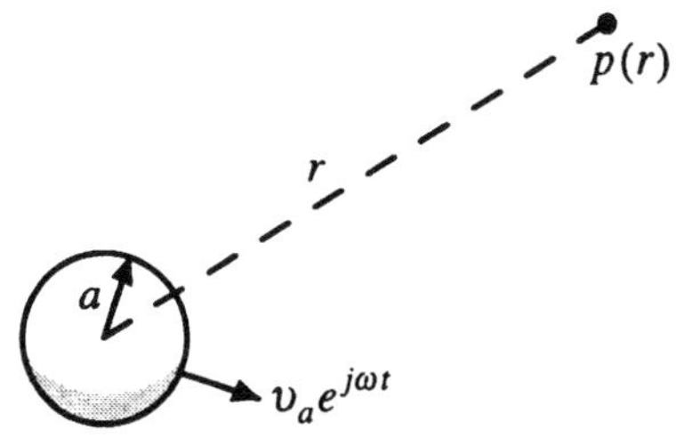

When the radius 'a' of the source is small compared to the wavelength of sound, this pulsating sphere becomes a "point source," or monopole. For such a source, the pressure wave produced is known to be proportional to the time rate of change of mass flow introduced by the source, and as explained above, inversely dependent on the dis-tance from the source. This leads to a formula for the pressure at any radius $|p_{rad}| = \omega \rho_0 A_S v_s / 4\pi r$, where the source area $A_S = 4\pi a^2$ and because the sound power is

$$\Pi_{rad} = \frac{1}{2}|p_{rad}|^2 \frac{4\pi r^2}{\rho_0 c_0} = \left(\frac{2\pi a}{\lambda_{sound}}\right)^2 A_S \rho_0 c_0 \langle v_s^2 \rangle \qquad (8.4)$$

so by inspection, the quantity $(2\pi a / \lambda_{sound})^2$ is the radiation efficiency in this frequency range. Because it depends on the ratio of a dimension (the circumference of the spherical source) to the wavelength of sound, this is a simple example of a geometric radiation efficiency. However simple though, it illustrates a general property of the geometric radiation efficiency, that when the dimension of the radiator becomes greater than the wavelength of sound, this factor becomes unity. In this case, that dimension is the circumference of the sphere, which is a good guide to selecting a characteristic dimension for irregularly shaped structural components like I beams; wrap a string around the object and record its length, that is its characteristic dimension as far as sound is concerned.

The low-frequency and high-frequency regions of σ_{geom} are joined as shown in Figure 8.8. Because the low-frequency behavior is proportional to ω^2, this radiation efficiency rises at 6 dB/octave toward the "corner frequency" where $\lambda_{sound} = 2\pi a$. From the shape of this curve near and above the corner frequency, we can see that an approximate value for the effective dimension will have only a small effect on an estimate of σ_{geom}, but at lower frequencies a poor estimate can have an effect on the accuracy of an estimate. These considerations are important because one frequently must use such estimates to get an idea of how a vibration is to be interpreted for its potential to radiate sound. Because we know both the dimensions of the product we are dealing with and the wavelength of sound as a function of frequency, we can often estimate the geometric radiation efficiency by inspection.

Figure 8.8   **The geometric radiation efficiency for various one- and two-dimensional (plates and beams) for pulsating and translating types of motion.**

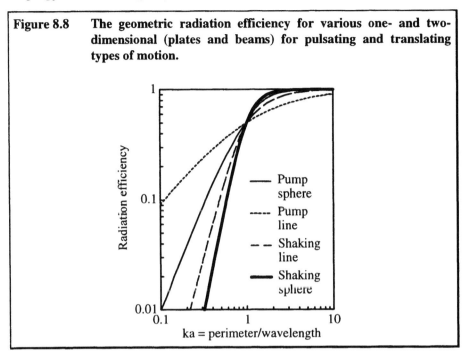

We noted above that the sound pressure radiation from a pulsating source is the time derivative of the mass flow rate from that source. The example shown in Figure 8.9 is the exit flow from the muzzle of a gun, where a certain amount of mass exits the muzzle in the time τ_s. The mass flow rate and its time derivative are shown. In this case, the pressure waveform is a positive pulse followed by a negative one, a so-called N-wave. All general sources of this type—explosions, bursting balloons, and gunfire—produce N-waves as their acoustical signature.

8.4
THE VIBRATING DISC; THE ACOUSTIC DIPOLE

There are structures that do pulsate, and pulsating airflow from a pipe also radiates like the pulsating sphere as long as the circumference of its opening is small compared to λ_{sound}. A more common situation is when a structure vibrates (shakes) and displaces air from one side while drawing it in on the other. This situation can be modeled as shown in Figure 8.10 showing a vibrating disc, which when small compared to λ_{sound}, it is called an acoustical dipole. The total force to make the disc vibrate is partly due to the mass of the disc and an accretion mass (or added mass) due to the air reaction. The force required to accelerate the added mass of the air is the source of sound radiation, and the radiated sound pressure is proportional to the time derivative of this force

$$|p_{rad}| = \frac{\omega F_{obs}}{4\pi c_0 r} \tag{8.5}$$

where F_{obs} is the component of force applied to the air, in the direction toward the observer. Because the force is proportional to the acceleration of the disc, then $p_{rad} \propto \omega^2 v$ and the radiation efficiency of the dipole varies as ω^4, or 12 dB/octave. Detailed analysis shows that the "corner frequency" in this case is the same as for the sphere if 'a' is the radius of the disc. Above the corner frequency, the radiation efficiency of the vibrating disc again approached unity.

Figure 8.9   **The sound from a monopole source that injects material into the air has "N-wave" time and space dependence.**

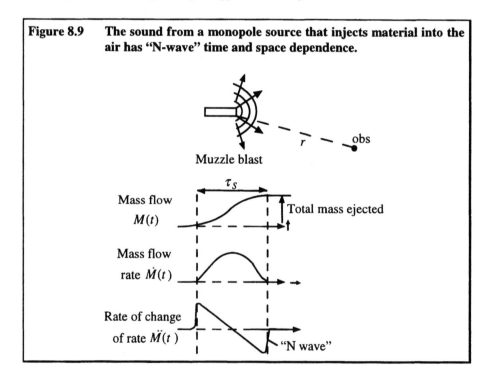

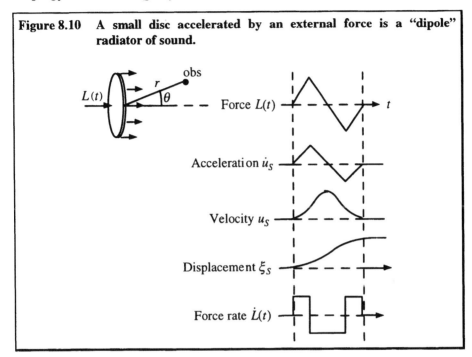

Figure 8.10 A small disc accelerated by an external force is a "dipole" radiator of sound.

The dipole model is useful in explaining how oscillating or decelerating objects radiate sound (a golf club for example. The dependence of the radiation efficiency on a higher power of frequency indicates that a dipole does not radiate sound as well as a monopole does at low frequencies. That is why a bass speaker will not radiate sound very well if one removes the back from the speaker cabinet. But the dipole model is equally important in showing why bodies that *do not vibrate* still radiate sound when placed in a flow as noted in Chapter 5.

When a rigid body is placed in a flow, turbulence that impinges on the body and turbulence that is shed by the body both cause forces to be exerted by the flow onto the body and, by reaction, by the body onto the flow. This reaction force produces sound in the air just as the force diagrammed in Figure 8.10. Another way to see the situation is to imagine a fixed region in a flow (shown in Figure 8.11). In this case, any sound radiation must come from the free turbulence, which will be very weak. In order for the air in that region to remain fixed, external body forces must be applied to the region, with both constant and fluctuating components. It is the fluctuating parts of this force that are the forces that the actual rigid body applies to the flow around it that produce the sound.

Figure 8.11   **A rigid body in turbulent flow is a dipole radiator of sound. The flow at any location can be thought of as a combination of free turbulent airflow (weak radiation) and a vibrating sphere (to make the total motion of the fluid vanish). The vibrating sphere can be thought of as radiating the sound.**

(a) Turbulent flow region with virtual body

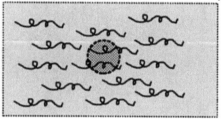

(b) Turbulent flow region with rigid body

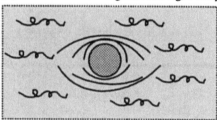

8.5

COMBINED MONOPOLE AND DIPOLE RADIATION; EXAMPLE OF A GOLF BALL

We have seen that an acoustical monopole is related to volume displacement and a dipole is related to rigid translation. Both events occur when a ball is struck. A well-examined case is the golf ball that is known to be rapidly accelerated when struck, remaining in contact with a driver only about one-half msec. A series of sketches of the impact is shown in Figure 8.12. It is clear from these sketches that the acceleration of the center of mass and ball compression both take place during the impact.

Figure 8.12   **A golf ball when struck is both compressed (volume pumping) and accelerated, so it acts as both monopole and dipole. Total contact time is about one-half msec.**

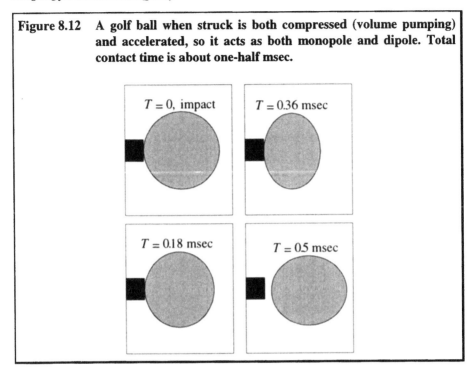

The compression is a monopole that radiates uniformly in all directions, whereas the sound due to translation (and acceleration) of the ball has a directivity. There is no radiation perpendicular to the direction of impact for this part, so if we measure the sound to the side of the driver where only the compression radiates, and from behind where both radiate, we can see the relative radiation of both components.

Time-frequency graphs of the sound at these two locations are shown in Figures 8.13a and b. Figure 8.13a is the spectrogram of the sound from the rear microphone, with a band of high radiation below 4 kHz and some peaks of radiation at higher frequencies. The frequency range above 4 kHz is dominated by resonances of the club, and the broad hump of energy below 4 kHz is due to compression and acceleration of the ball. This frequency range is determined by the circumference of the ball, which is a wavelength at 4 kHz. Note that the spectrogram has a duration of about 0.1 second, much longer than the period of impact. This spreading of the signal in time is due to the frequency analysis. When a signal is analyzed in narrow frequency bands, it is extended in time by an amount equal to the reciprocal of the bandwidth.

Figure 8.13   (a) The spectrogram of golf ball sound at a location along the line of impact shows energy below 4 kHz and resonance features above 5 kHz. At this location both compression and acceleration of the ball contribute to the sound.

(b) Below 4 kHz, the sound spectrogram at a location perpendicular to the line of impact is due to compression of the ball.

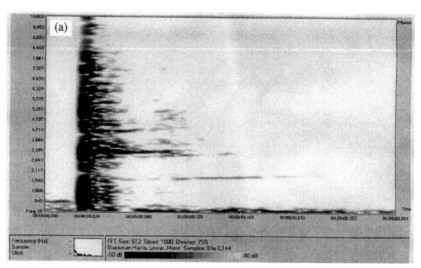

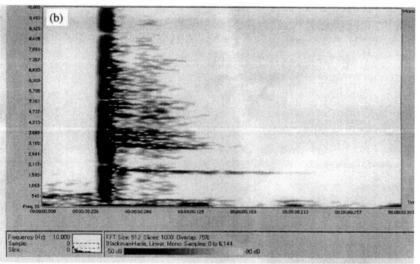

The spectrogram for sound radiated to the side is shown in Figure 8.13b. There is substantially less sound radiated to the side below 4 kHz indicating that the monopole sound due to ball compression and the dipole sound due to acceleration are both important in the impact sound.

The actual spectrum of the sound, from Equation 8.2 is a product of the spectrum of the velocity and the radiation efficiency. The sound of the ball acceleration involves a step in velocity from rest to final velocity in 500 μsec, which has a spectrum that is proportional to f^{-2} up to a frequency

$$f_{max} = 1/2\pi\tau_{rise} = 2000/2\pi = 300 \text{ Hz},$$

which is an estimate but certainly reasonable. Above 300 Hz, the velocity spectrum will vary as f^{-4}. Because the dipole radiation efficiency varies as shown in Figure 8.14, then we see that the combined effects of the velocity and radiation efficiency produces a spectrum of radiated sound that is flat from about 500 Hz to 4 kHz.

The ball compression velocity is a pulse because the volume compression is a step. The pulse has a flat spectrum up to 300 Hz, dropping off as f^{-2} above. When combined with the radiation efficiency spectrum, we get the spectrum shown in Figure 8.14b, a spectrum that is similar to that of the dipole sound. These examples show how one can combine the spectra of velocity and radiation efficiency to predict the spectrum of radiated sound.

8.6
THE ACOUSTIC QUADRUPOLE; BILLIARD BALL EXAMPLE

The click of two colliding billiard balls is the result of a pair of impact forces acting in opposite directions on the two balls, essentially two dipole sources in opposite directions at slightly different places. The situation is sketched in Figure 8.15. Immediately before the collision, the air is being displaced and drawn in as shown in (a). Immediately after the collision the airflow has reversed as shown in (b). This step in velocity creates the sound of the collision, and the combination of the two "back-to-back" dipoles is called a *longitudinal quadrupole*.

Figure 8.14 **(a) The velocity spectrum of the ball CG is combined with the dipole radiation efficiency to predict the sound spectrum for ball acceleration.**

(b) The velocity spectrum of ball compression is combined with the monopole radiation efficiency to predict the sound spectrum for ball compression.

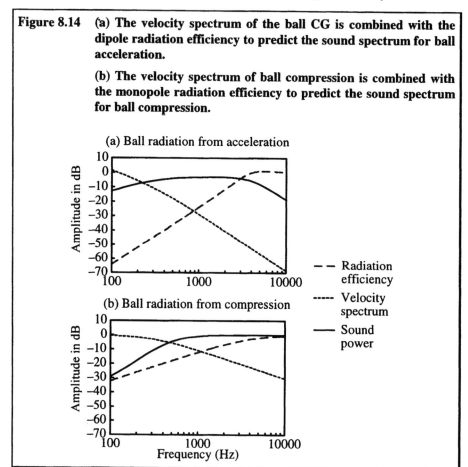

Figure 8.15 **Two balls colliding produce a click due to the sudden reversal of airflow around the balls. This is a "quadrupole" sound and for billiard balls is not related to resonant vibration of the balls.**

The situation in Figure 8.15a and b is also appropriate to the case of a ball hitting a rigid surface, as in (c), because the image of the ball acts acoustically as the second ball. Because the two balls are so closely spaced, the sound that is generated is now proportional to the *second* derivative of the velocity, producing a waveform for the collision like that shown in Figure 8.16. Note that the sound is not due to the vibrational ringing of the balls as they impact, but rather to the abrupt change in airflow around the balls as they change their directions of motion. In the data of Figure 8.16, an increase in amplitude of the data by a factor of 50 does show the ringing of the balls, but this is not the sound one hears as a "click."

When parts collide we would therefore expect to produce an impact sound radiated as quadrupole sound. If we think of the longitudinal quadrupole as a pair of oppositely directed forces acting on the air, then it is clear that time-dependent stresses within turbulent airflow will produce such paired forces. Quadrupole radiation from such stresses in a flow (called Reynolds stresses) are the causes of radiated sound from turbulent flow. But, aside from airflow from high-speed air jets, this source of sound is generally weak compared to other sources of flow noise like wake shedding or airflow impingement that involve forces on rigid objects.

Figure 8.16   **The observed sound pulse from colliding balls has the expected waveform for a quadrupole. Resonant (and inaudible) vibrations are apparent if the gain on the signal is increased.**

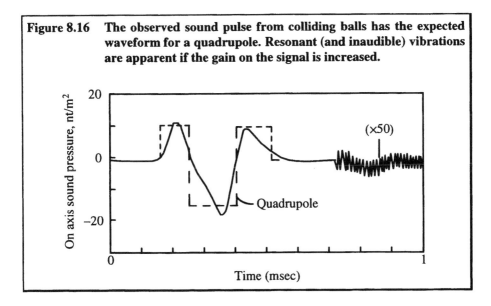

8.7
COMBINING MONOPOLES TO PREDICT SOUND RADIATION; THE BOUND-
ARY ELEMENT METHOD (BEM)

A finite element calculation of structural vibration in response to the forces that drive
it will usually end up with a model for the vibration of the node points, or by a
reordering of the data, the vibration of the mesh elements bounded by the nodes, as
shown in Figure 8.17. Alternately, the vibrational velocities might be known based on
measurements on the product. If there were an array of point sources in free space,
adjusted in strength so that the total volume velocity through each element is the same
as for the actual structure, then the sound from those sources should approximate the
sound from the actual structure.

   This discussion of BEM basics should not be construed as an indication of how
commercial software packages actually work, for two reasons. First, it is possible to
formulate the radiation problem in more than one way. Second, the straightforward
approach taken here can lead to difficulties in certain situations that are resolved by
adding on extra features to the algorithms. The following should therefore be looked
at as a way of understanding the basic idea of BEM techniques, and not as a recipe for
software development.

Figure 8.17   **The surface elements of a product are replaced by elementary
sources of sound that result in the same volume flow across each
element as in the product. The calculation using such a method is
an acoustical boundary element method.**

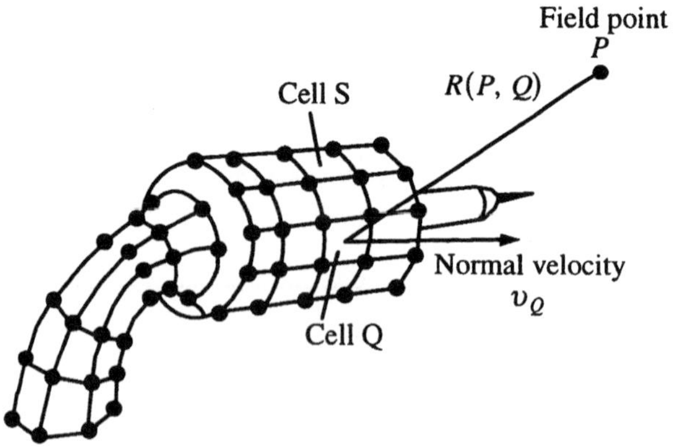

If we consider two structural elements labeled 'S' and 'Q' in Figure 8.17 and drawn separately in Figure 8.18, the volume velocity source U_0 at 'Q' will create an average velocity outward

$$v_Q = U_Q/2A_Q,$$

whereas a source U_S at 'S' will produce an outward component of velocity at 'Q' to be added to that due to the local source at 'Q'. The sum of all the effects of the sources 'S' $\neq$ 'Q', added to the local source must then equal the actual normal velocity at the element 'Q.' This then leads to an expression for the needed volume velocities

$$\frac{U_S}{2A_S} = \sum_{Q \neq S}(T_{QS} + \delta_{QS})^{-1}v_Q^{\mathrm{meas}} \tag{8.6}$$

where T is a transformation matrix that depends on the geometry of the structure, the speed of sound, and the frequency (essentially the ratio $D/\lambda_{\mathrm{sound}}$):

$$T_{QS} = \frac{(jk)^2}{4\pi\gamma_{QS}}\left(1 + \frac{1}{\gamma_{QS}}\right)e^{-\gamma_{QS}}\cos\theta_{QS} \tag{8.7}$$

where $\gamma_{QS} = jkR_{QS}$.

Figure 8.18   **Diagram showing how elements interact to produce the desired velocity of a surface (velocity due to local source plus velocities induced from all other sources).**

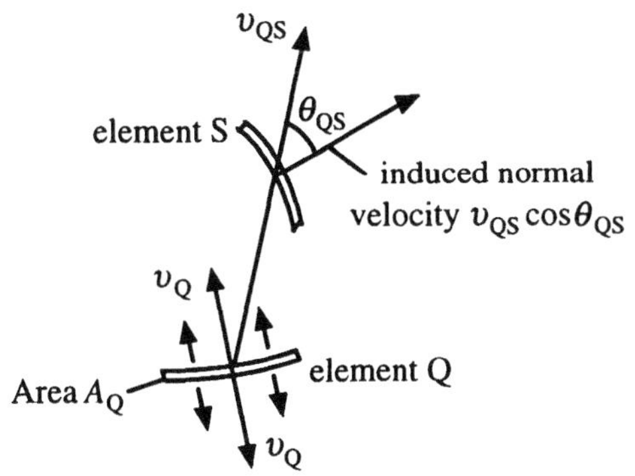

Once the volume velocities are calculated from the measured or calculated mesh element velocities according to Equation 8.6, then the radiated sound pressure is readily computed at an observation location 'R' according to

$$p(R) = \sum_Q \frac{j\omega\rho_0 U_Q}{4\pi \, r_{QR}} \, e^{-j\omega r_{QR}/c_0} \equiv \sum_Q \Gamma_{QR} U_Q \tag{8.8}$$

which can be combined with Equation 8.6 to form a single linear operation on the mesh element velocities to calculate the radiated pressure.

8.8
SUMMARY

The estimation of sound power from a vibrating structure can be thought of as a determination of its radiation efficiency. The radiation efficiency depends on two ratios of length scales, the ratio of the dimension of the product to the wavelength of sound (the geometric radiation efficiency) and the ratio of the wavelength of structural vibration to the wavelength of sound.

Elementary sources that displace volume or produce forces on the air have radiation efficiencies determined by their size. These sources arise from combustion, impacts, and aerodynamic loads. The sound from elementary sources can be combined in a computational method to predict sound from more complex patterns of vibration in so-called boundary element methods (BEM).

CHAPTER **9**

Sound Radiation by Products and Its Measurement

This chapter considers sound radiation from the vibrating surfaces of a product. The principal parameter to be determined is the dynamic radiation efficiency, introduced in the last chapter. The dependence of this parameter on structural geometry, materials, and assembly is a prime element in design for SQ.

The measurement of product sound in the laboratory and in production can be accomplished in various ways. Here we consider three methods: acoustical intensity, and hemianechoic and reverberant room methods.

9.1
RADIATION BY FAST WAVES; DYNAMIC RADIATION EFFICIENCY

We now consider cases where the structure is large compared to the wavelength of sound, and the geometric component of radiation efficiency is not a factor. Now the scale of the vibrations, expressed as the ratio of a wavelength of the vibration to the wavelength of sound $\lambda_{\text{vib}}/\lambda_{\text{sound}} = c_{\text{sound}}/c_{\text{vib}}$, determines the radiation efficiency. We have called this factor in the radiation efficiency the *dynamic radiation efficiency*.

As discussed in Chapter 7, bending waves are usually of the greatest interest for acoustical radiation and excitation because sound pressures excite them directly, and they displace air directly when the plate vibrates. We also saw that the speed of bending waves changes with frequency. As a result, there will usually be a frequency at which the bending wave speed exactly equals the speed of sound in the air; $c_b = c_0$. The frequency at which this occurs is the *critical frequency*, which is a very important parameter. Using the formula for the bending wave speed in section 6.2, the critical frequency for bending waves on a steel or aluminum plate is

$$f_c = 13000/h \, (\text{mm}) \tag{9.1}$$

For example, an aluminum panel 3 mm thick has a critical frequency of 4.3 kHz. If $f > f_c$, and therefore $\lambda_{\text{vib}} > \lambda_{\text{sound}}$, the bending wave is supersonic, and if $f < f_c$ and $\lambda_{\text{vib}} < \lambda_{\text{sound}}$, then the bending wave is subsonic. This will usually make a big difference in how well the vibrations of the structure radiate sound.

Figure 9.1 illustrates how a supersonic disturbance (a projectile in this example) produces a sound wave that aligns itself so the sound can keep up with the disturbance. The angle at which the wave travels so this is possible is called the *Mach angle*, named after the Austrian aerodynamicist Ernst Mach. A similar effect occurs when a supersonic bending wave travels along a plate as shown in Figure 9.2. The sound wave is radiated at the Mach angle $\theta = \sin^{-1}(c_0/c_b)$, so that for the particle velocity in the wave $v_{\text{wave}} = p/\rho_0 c_0$ to have a component perpendicular to the plate that matches the vibrational velocity of the plate $v = v_{\text{wave}} \cos\theta$, the pressure in the sound wave must be $p = v \times \rho_0 c_0/\cos\theta$. This means that the effective dynamic radiation efficiency is

$$\sigma_{\text{dyn}}(f > f_c) = 1/\cos\theta = (1 - \lambda_{\text{sound}}/\lambda_{\text{vib}})^{-1/2} = (1 - f/f_c)^{-1/2}. \tag{9.2}$$

Figure 9.1   **A supersonic projectile produces sound waves traveling at the Mach angle, which allows the sound wave to "keep up with" the disturbance.**

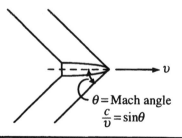

$\theta = $ Mach angle
$\dfrac{c}{v} = \sin\theta$

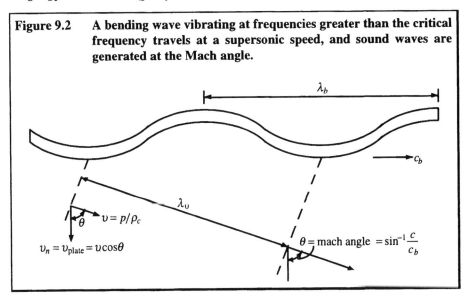

Figure 9.2   A bending wave vibrating at frequencies greater than the critical frequency travels at a supersonic speed, and sound waves are generated at the Mach angle.

This radiation efficiency diverges at the critical frequency (benignly, it is finite when averaged over a frequency band), but that is a theoretical result that does not occur in practice. There is usually a peak in the radiation efficiency just above the critical frequency due to this effect, however. This function is graphed in Figure 9.3.

9.2
RADIATION AT FREQUENCIES BELOW THE CRITICAL FREQUENCY; SOUND FROM (ACOUSTICALLY) SLOW WAVES

The simple notion of sound radiation in the preceding section would predict no sound radiation at all for frequencies below the critical frequency; $f < f_c$. But of course there is radiation at these frequencies. A boundary of or an obstruction on the plate will cause wave reflections and a standing wave on the plate. Because $\lambda_{\text{sound}}/\lambda_{\text{vib}} > 1$ in this frequency range, there will be local "sloshing" of the air from peak to trough in the standing weave over most of the plate surface. This occurs everywhere except along the boundary or obstruction, where there is a net displacement of volume velocity. It can be shown that this uncanceled volume velocity is proportional to the force that the obstruction exerts on the plate. So we get the paradoxical result that the sound is being radiated from regions of the structure where obstructions are resisting its vibration!

Figure 9.3   **The radiation efficiency for a simply supported plate has distinctly different forms above and below the critical frequency.**

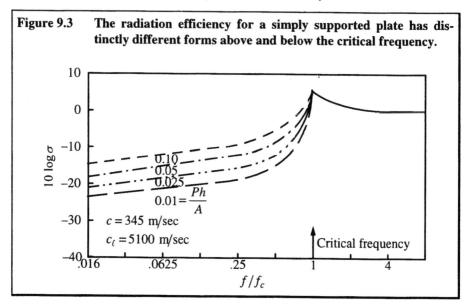

Just as the wavelength of the vibrations above the critical frequency must be greater than the wavelength of sound for sound radiation to occur, the wavelength *along* the obstruction must be greater than the wavelength of sound for sound to be radiated below the critical frequency. For this reason, only a fraction of the bending waves (or of the resonant modes) will radiate sound. The two factors of the radiation from the edge and the fraction of modes that radiate combine to produce the dynamic radiation efficiency for $f < f_c$:

$$\sigma_{\text{dyn}}(f < f_c) = \frac{2}{\pi^2} \frac{Ph}{A} \sin^{-1}(\frac{f}{f_c})^{1/2} \tag{9.3}$$

where P is the edge length (perimeter) of the plate, h is its thickness, and A is its area. This function is graphed in the frequency range $f < f_c$.

There are a number of detailed assumptions that go into the derivation of Equation 9.3 that affect details of the curve in Figure 9.3 for $f < f_c$, but the general features of the curve apply in most practical situations. The radiation efficiency drop below the critical frequency will occur, and radiation in this frequency region will increase as the length of obstructions P is increased. This is one reason why adding stiffeners to a plate structure will usually *increase* sound radiation, contrary to what we might expect.

There is another way to interpret this discussion. It can be shown that any force on the plate results in a deflection that displaces volume velocity. Edge supports, but also obstructions, stiffeners, and excitations from mechanisms in the product like those discussed in preceding chapters all produce forces, and these cause radiation. For example, a force due to a point obstacle on a plate below the critical frequency radiates sound power

$$\Pi_{rad} = \langle f^2 \rangle_t \frac{\rho_0}{2\pi \rho_s^2 c_0} = \langle v_p^2 \rangle_{x.t} \left| \frac{1}{G_p + Y_{int}} \right|^2 \frac{\rho_0 c_0}{2\pi \rho_s^2 c_0^2}, \tag{9.4}$$

which leads to a radiation efficiency

$$\sigma_{dyn} = \left| \frac{1}{G_p + Y_{int}} \right|^2 \frac{A}{2\pi G_0^2 m_p^2} \tag{9.5}$$

which depends on the plate conductance and the internal mobility of the obstruction. Interestingly, the calculation of forces from edges and obstructions on a plate and the use of Equation 9.4 leads to a formula for a radiation resistance and radiation efficiency identical to Equation 9.3.

9.3
MODIFYING STRUCTURAL SOUND RADIATION

We have noted in Chapter 8 that radiated sound power is a product of ms velocity and radiation resistance. The latter has also been expressed by the radiation efficiency. Once the vibrational noise energy has been transmitted into a structure and a ms vibrational velocity has been established, then the modification of sound radiation can only come by decoupling the structural vibration from the surrounding air.

One way to do this is to use an add-on decoupling treatment, usually consisting of a resilient layer of material combined with a layer of mass, as sketched in Figure 9.4. This combination of mass and spring acts as an isolator between the vibrating structure and the air, reducing the sound radiated by the structure. It is also possible to incorporate a damping treatment into this assembly as shown, thereby reducing the resonant (damping controlled) vibration amplitudes in addition to reducing the coupling to the air.

Figure 9.4   An add-on treatment of a resilient layer and a limp mass septum acts as a decoupler between panel vibrations and the surrounding air.

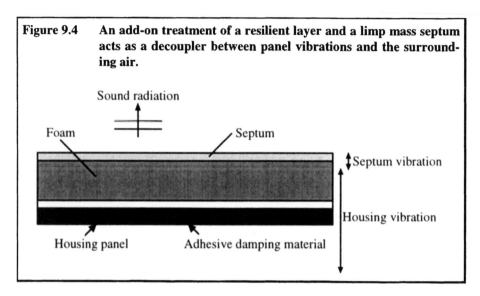

Another and often very effective way of reducing the radiated sound from a structure can be illustrated using the copier drive system discussed in Chapter 4. We discussed the excitation and source of the noise energy as due to chains and sprockets and gear meshing, but these components, being small, do not radiate much sound energy. In that example, the supporting casting for the drive and idler sprockets, the motor, and the gears is the structural casting shown in Figure 4.13. This reinforced plate structure has the role of rigidly locating the drive components, but because it is stiff and solid, it is also a very good radiator of sound. The measured radiation efficiency of this structure is shown in Figure 9.5a. We see that it becomes near unity at frequencies greater than 1 kHz.

In this case, there is no reason that this structure must be solid, because the structure is completely internal to the product, and it does not need to act as a physical barrier for any purpose. It must only be rigid enough to support the mechanisms attached to it. If we cut away the webs between the reinforcing beams in the structure, as shown in Figure 9.5b, we end up with a truss structure that is only slightly less rigid than the original structure. The original stiffness can be recaptured if needed by slightly increasing the thickness of the truss elements. But the measured radiation efficiency is greatly reduced in the frequency range below 2 kHz.

Figure 9.5 **A casting that supports a copier mechanism has**

(a) much lower radiation efficiency at lower frequencies, when

(b) the webs between ribs are removed to change it from a reinforced plate to a truss.

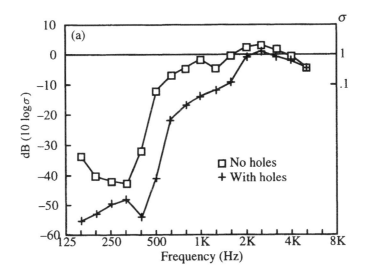

The drop in radiation efficiency below 2 kHz very nicely complements the drop in noise energy above 1500 Hz that was achieved by modifying the sprockets. When a solid plate structure can be replaced by a truss structure or a perforated plate, significant reductions in radiated sound can often be obtained quite simply. We have shown in Chapter 5 that the sound radiated by a brushless DC motor can be significantly reduced by perforating the motor housing structure without a deleterious effect on motor performance. Perforations or replacement of solid plates by expanded metal can be acceptable if the plate structure is not containing a fluid or protesting against dust.

9.4
THE EFFECTS OF CURVATURE ON SOUND RADIATION

When a structure is curved, it is stiffened in the direction of curvature, and that has a very significant effect on sound radiation. We noted in Equation 8.2 that sound radiation depends on the product of response (ms velocity) and coupling to the sound field (radiation efficiency). Curvature can affect both of these factors.

The ms velocity is determined by force excitation, mechanical conductance, and damping. If the other factors are fixed, the conductance determines the ms velocity. The conductance is proportional to modal density, which is in turn affected by curvature. This is best illustrated by comparing the response of a cylinder to that of the flat plate from which it is formed. Modes of the plate that try to bend the cylinder along its axis are stiffened by membrane stresses in the cylinder, causing them to resonate at a higher frequency.

Membrane stresses in the cylinder travel at the longitudinal wave speed, and therefore will not stiffen the cylinder if they do not have enough time to travel around the cylinder

$$T_{\text{mem}} = 2\pi\, a/c_\ell = 1/f_{\text{ring}} \,,$$

where f_{ring} is the ring frequency. For frequencies $f > f_{\text{ring}}$, the membrane stresses are ineffective, and the modes and modal density of the cylinder are unaffected by curvature. The modes of the cylinder that are stiffened at frequencies $f < f_{\text{ring}}$ "pile up" at f_{ring}, producing a modal density like that shown in Figure 9.6. The peak in density at the ring frequency is one factor affecting sound radiation through its effect on the amplitude of vibration.

Figure 9.6 **A cylinder has reduced modal density below the ring frequency because of membrane stresses that make the structure stiffer in its axial direction. When these axial modes are "released" at the ring frequency, they produce an increase in the modal density.**

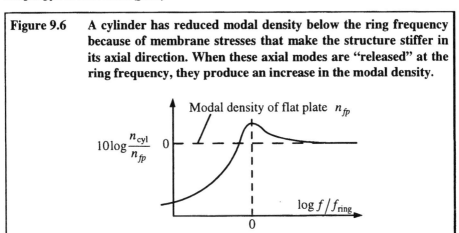

The ring frequency for steel or aluminum cylinders is

$$f_{ring} - 16000/d(in)$$

(steel or aluminum), so that a product with a cylindrical housing of 16-inch diameter has a ring frequency of 1 kHz. If the material is plastic, this frequency is reduced by about an octave. As noted above, modes that would resonate at lower frequencies without curvature will resonate at higher frequencies with curvature. These modes lose the stiffening effect at the ring frequency and resonate near the ring frequency. We therefore expect extra modes in this frequency range for cylindrical structures and increased response.

In addition to the modal density effect, some of the modes that do resonate near the ring frequency have wavelengths that are greater than the wavelength of sound, making them "acoustically fast." This effect leads to an increase in the radiation efficiency at the ring frequency for cylinders that have a frequency less than the critical frequency. If the critical frequency is also less than the ring frequency, the radiation efficiency is already at or above unity. This effect on radiation efficiency is sketched in Figure 9.7. Because the critical frequency in Equation 9.1 can also be written

$$f_c = 500/h(in),$$

the ring frequency will be less than the critical frequency when $h/d < 1/32$, which is a fairly common situation for products.

Figure 9.7   **A comparison of the radiation efficiency of a cylinder and a flat plate of the same area shows a strong peak at the ring frequency for the cylinder when the ring frequency is less than the critical frequency. Some of the cylinder modes released at the ring frequency have supersonic axial wave speeds.**

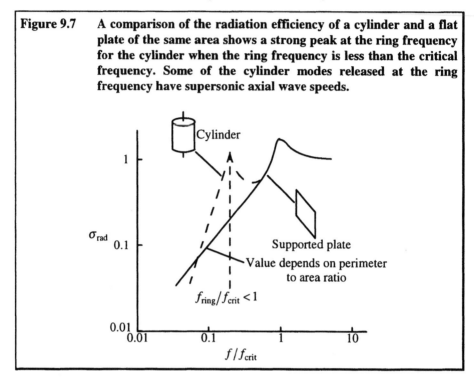

9.5
SUMMARIZING DYNAMIC RADIATION EFFICIENCY

From this discussion, we can gather some main points. If we can estimate the forces on a structure, we can first make an estimate of sound radiation assuming the structure is a flat plate of the same area. Obstructions and supports increase the radiation efficiency for $f < f_c$ according to Equations 9.3 and 9.5. If the structure is cylindrical, the response is modified by the modal density factor in Figure 9.6, and the radiation efficiency is modified by the factor in Figure 9.7.

Further, at low frequencies when the wavelength of sound becomes greater than a characteristic dimension of the structure, the additional geometric radiation efficiency factor shown in Figure 7.8 should be applied. Although these relations and predictions are not exact, they will often provide quite a good approximation to the measured results, and they have the distinct advantage of retaining parametric dependence on the various dimensions and material properties of the product so that one can anticipate how the sound radiation will change as modifications are made to the product.

9.6
PRODUCT SOUND MEASUREMENT METHODS

The engineering function described in Figure 1.2 requires measurements for product analysis and design. These measurements can determine sound pressures at particular locations, total radiated sound power, the radiation efficiency of the structure, and the relative contribution to radiated sound from various mechanisms in the product. They are characterized by dedicated spaces—reverberation rooms, semianechoic rooms, and anechoic chambers—and instrumentation that is flexible and adaptable to new situations and product changes.

There is also a need for product sound measurements on the production line to confirm performance or conformity to a standard. These measurements can simply provide a pass-fail decision for the item, or they can identify the fault to determine how the production procedures should be changed to avoid failures. The space for the measurement, the arrangement of sensors, and the processing of data are all likely to be different from the situation in the design laboratory.

In the following sections, we discuss three commonly used methods for determining sound power: sound intensity, pressure scans, and reverberation rooms. These methods, implemented in different ways for design and production, have to be adapted to the specific needs for a product. Because we cannot deal with more than a few situations, we will discuss just how the methods we describe have been adopted for the situations described.

9.7
SOUND INTENSITY IN A PRODUCT DESIGN FUNCTION; COMPARISON WITH PRESSURE SCAN

The sketch in Figure 9.8 shows a typical measurement layout in a hemianechoic room (hard floors, highly absorbing walls and ceiling). A hemispherical surface is imagined around the product, and a set of locations is defined, each location corresponding to an area on the surface. Generally, locations will be chosen so they represent an area that is approximately equal to the others, but that is not necessary and may not be desirable if the product tends to radiate sounds much more in some directions than others. In that case, measurement points will be taken closer together along those directions.

Figure 9.8   **A measurement of sound radiated by a product in a hemianechoic chamber at 2 radii and using both pressure and sound
intensity methods.**

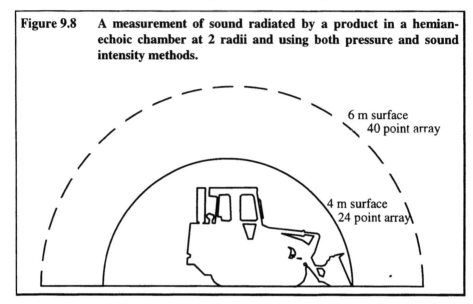

6 m surface
40 point array

4 m surface
24 point array

One purpose of the measurement is to determine overall sound power

$$\Pi_{rad} = \sum_n I_n A_n \qquad (9.6)$$

where A_n is the area assigned to each measurement, and each I_n is the sound intensity
associated with that area. In a reflection free environment, sound might be assumed to
be propagating radially outward from the product so that the intensity is determined
by the sound pressure

$$I_n = \langle p_n^2 \rangle_t / \rho_0 c_0 . \qquad (9.7)$$

This estimate of the intensity relies heavily on the assumption that the waves are freely
propagating (no reflections from the walls) *and* that they are directed radially outward
from the center of the sphere. The sound power spectrum determined this way for the
front loader shown in Figure 9.8 is presented in Figure 9.9. One can also estimate the
directivity of the source using this data because the directivity in any direction is
defined as

$$D_n \equiv I_n / \langle I \rangle = I_n \left(\sum_n A_n \right) / \Pi_{rad} \qquad (9.8)$$

Figure 9.9 **The spectra of average pressure or intensity levels for the two methods indicates that the pressure method is about 1.5 dB higher than the intensity method, perhaps because the sound does not progress outward from the product in a purely radial direction.**

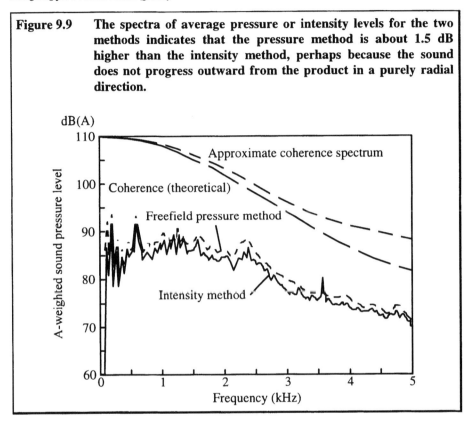

For many years this was the only way to determine sound power and directivity of a product, but digital processing has made another method available. A *sound intensity* measurement uses a pair of microphones to determine both sound pressure and particle velocity, because it is the time average product of these two variables that is the intensity, and the intensity is in the direction of the particle velocity. In a freely propagating wave, the particle velocity $v = p/\rho_0 c_0$, leads to Equation 9.7. But if the sound field has reflections or is not propagating normal to the measurement surface, using Equation 9.7 will introduce an error.

With two microphones closely spaced, one can average their outputs to determine sound pressure and take their difference to determine the acceleration of air particles that is readily converted to velocity by dividing by $j\omega$. Very small phase errors in this process can introduce large errors in the estimate of intensity, which is one reason why digital processing was necessary for its success. The calculation results in this formula for the sound intensity[17]

$$I_n = \text{Im}(S_{ab})/2\omega\rho_0\Delta x \qquad (9.9)$$

where Δx is the spacing between the microphones that record the two pressures p_a and p_b. The quantity $S_{ab} = \langle P_a P_b^* \rangle$ is the cross spectrum of the two pressure signals. Although it is possible to carry out these calculations with any two channel digital analyzer, many engineers find it useful to obtain special sound intensity options that have helpful features, including mapping software.

The graph in Figure 9.9 shows the sound power determined using a sound intensity measurement in addition to the pressure measurement already discussed. We see that the results are very similar, except that the pressure measurement results in sound power that is about 1.5 dB greater. Presumably this is because the sound is not traveling in a purely radial direction outward from the center of the sphere. Because a sound wave traveling at an angle with respect to the measurement surface represents less energy flow through the surface, the pressure measurement, which doesn't account for the direction the wave is going, will overpredict the intensity.

When the measuring surface surrounds the product as in Figure 9.8, if the source is steady, the total power is (theoretically) not affected by the sound from other sources that both enters and leaves the measurement surface. It is sometimes considered that this property will allow sound intensity to be used as a production line measurement where other sources are present. But the ability to reject other sources is very limited, and it would be unwise to rely on such rejection in a production line environment.

9.8
REVERBERATION ROOM METHODS

Reverberation rooms are usually thought of as strictly laboratory facilities, but as we will show, they can also serve as production line test chambers under certain circumstances. The basic idea of a reverberation room is to capture the radiated sound power from a product and convert that power into space averaged sound energy that can be measured as a sound pressure. Figure 9.10 sketches the idea. Sound power

Π_{rad} is radiated from the source, and this energy continues to rebound in the space, but with each collision a fractional amount α is removed by wall absorption. The rate of removal is therefore equal to the power from the source after it is diffused by its first reflection from the walls:

$$\Pi_{abs} = E_{room} V_{coll} \alpha = \frac{\langle p_{rev}^2 \rangle}{\rho_0 c_0^2} V_{room} \frac{c_0}{d} \alpha = \Pi_{rad}(1 - \alpha) \tag{9.10}$$

where $d = 4V_{room}/A_{walls}$ is the "mean free path," the average distance a packet of sound energy will travel between collisions with the wall. This relation provides the needed relation between radiated sound power and ms sound pressure in a reverberation room:

$$\Pi_{rad} = \frac{\langle p_{rev}^2 \rangle A_{walls} \alpha}{4\rho_0 c_0 (1 - \alpha)} \equiv \beta \langle p_{rev}^2 \rangle. \tag{9.11}$$

One can calibrate the room as an instrument by measuring the absorption coefficient α or by placing a source of known sound power in the room and determining the parameter β directly.

Figure 9.10 The sound in a reverberation room creates a ms pressure field that is proportional to the radiated sound power and inversely proportional to room damping as measured by the "room constant."

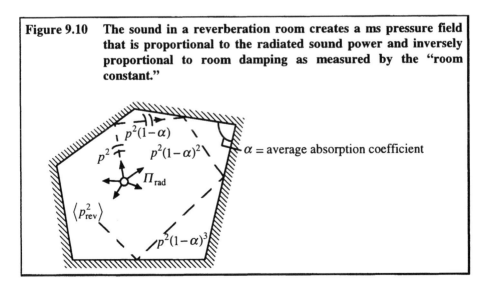

Figure 9.11 shows the interior of a laboratory reverberation room. The microphone is shown on a boom that rotates and sweeps it around the room to get a good ms pressure value. The panel on the wall provides some low-frequency absorption to increase the bandwidth of room modes as they become sparse at the lower end of the room's frequency range. The rotating vane periodically changes the shape of the room interior and provides an additional averaging effect of moving the source around.

There are many textbooks on the use and design of reverberation rooms. Here we will describe the use of a reverberation room for measuring the sound radiated by sewing machines on the production line. The example is the determination of A-weighted sound power for sewing machines. In this case, an examination of the spectrum of sound from the machine indicated that frequencies greater than 400 Hz determined the A-weighted spectrum, so that the reverberation room measurement could be restricted to that frequency range.

Figure 9.11   Basic elements of a reverberation room: a moving microphone, a rotating vane, and low-frequency sound absorbers.

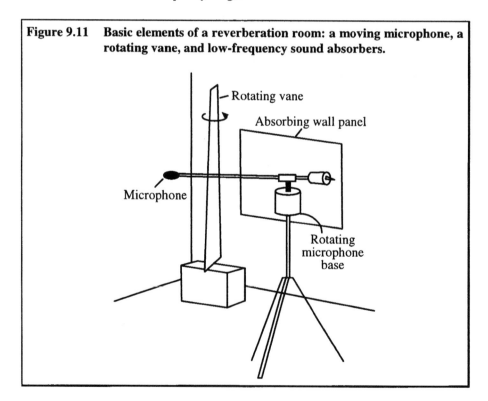

This means that the room could be much smaller than a conventional reverberation room, because the criterion for room size is the ratio of room mean free path to the wavelength of sound. Generally, the ratio will be about unity at the lowest frequency to be measured. A room designed to measure sound energy above 400 Hz needs to be only one-eighth in every dimension compared to one intended to measure down to 50 Hz. This room is to go on the production line, so it must be ruggedly constructed to be moved about on a forklift and have sufficient sound insulation so that factory noise does not disturb the measurement. Figure 9.12 shows the production line reverberation room.

The reverberant booth has a door that allows access for check out and calibration of the room, but during use this door is closed. The sewing machines are placed into the booth through a window as shown in Figure 9.13. The machine is placed on a rotating table that swings into the booth and places the machine so a photo cell can sense the motion of the needle bar for speed determination. The window is closed and the voltage to the machine is adjusted so that the machine comes to speed. If the machine will not achieve the speed within an acceptable voltage, it fails for that reason.

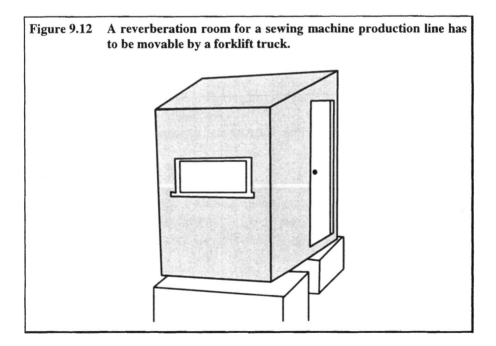

Figure 9.12   A reverberation room for a sewing machine production line has to be movable by a forklift truck.

Figure 9.13   The operator places the sewing machine into the room through a window; it is then tested for speed and radiated sound.

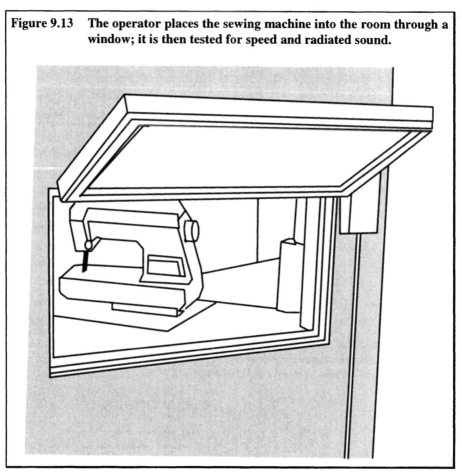

If the machine does come to speed, then the acoustical measurement commences. The sound pressure is measured by microphones at five locations within the space, three of which are shown in the diagram of Figure 9.14. These signals are squared and summed by the analyzer shown in the figure. If the sound level is in the required range (i.e., does not exceed the prescribed limit) then the machine passes the test. Passing this test means that the operator gets two green lights on the external panel shown in the figure. The entire test from placing the machine on the rotating table to placing the machine back on its production line pallet takes about one minute.

Figure 9.14 **A sketch of the production line reverberation room showing microphones, the low-frequency absorbers, and the signal analysis unit that determines machine speed and radiated sound.**

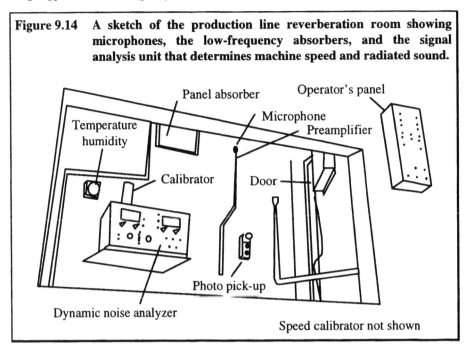

Panel absorber

Operator's panel

Temperature humidity

Microphone Preamplifier

Calibrator

Door

Photo pick-up

Dynamic noise analyzer

Speed calibrator not shown

If properly designed and used, this kind of measurement can be quite accurate. A cross plot of sound levels for 48 machines placed in the booth, compared to measurements of sound pressure in a hemianechoic room using five microphones at a distance of five ft. from the machine is shown in Figure 9.15. The scatter is less than 1 dB, and it was felt that in this case, the reverberant booth measurement was in fact more accurate than the hemianechoic room measurement, in part because only five microphone locations were used in the latter.

Although this reverberant booth was successful in making the desired measurement on the production line in an acceptable time, it has two characteristics that limit its usefulness. First, it has no diagnostic ability; machines are simply passed or failed. When a machine fails, other procedures have to be employed to determine why it failed. Second, A-weighted sound power, which correlates well with loudness, may not capture the more important perceptual aspects of the sound that make the sewing machine more or less acceptable to the user. We discussed these issues in Chapters 1 and 2 and have more to say about them in the next chapter.

Figure 9.15   **A cross plot of sound levels as measured by the production line reverberation room and a laboratory hemianechoic room shows good correlation between the two.**

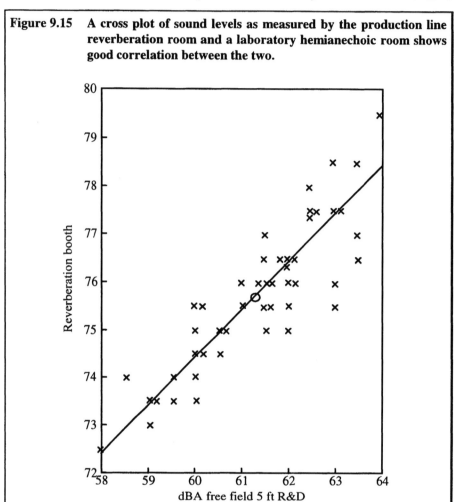

CHAPTER 10 Quality Assurance Using Vibration

In this chapter we return to the *Build* function of a product company. From a SQ point of view, the main concern is that the product as built performs as defined by Product Planning and as designed by Engineering. This may be done by monitoring either the product or process or both. The use of acoustical measurements for QA purposes was discussed in the last chapter, but most of the time a vibration test is likely to be preferred. After a general discussion of diagnostics and decision levels, two examples of the use of vibration are given.

10.1
THE PURPOSES OF PRODUCTION LINE TESTING

The basic purpose of production line testing for quality assurance (QA) is to avoid sending a defective product to the market, but within this purpose, other motivations exist that support the basic purpose. These include the determination of classes of faults (diagnostics of product and process) and prognostics and the determination of trends that show improvement or degradation of the production process. The purpose of this chapter is to provide some basic ideas about production QA, and using vibration signals generally and to illustrate the procedures with some examples.

The basic arrangement of a diagnostic system is indicated in Figure 10.1. Signals from the product are picked up by one or more sensors, and these signals undergo "conditioning," which means filtering, amplification, and change of impedance level. Some sensors, like accelerometers, have a very high internal impedance, which makes the signals from them susceptible to contamination by stray voltages or magnetic fields. A "charge to voltage converter," sometimes within the accelerometer case itself, changes source impedance to a much lower level. This device may also provide amplification and/or filtering of the signal.

Because most of the processing and interpretation of the signals will be digital, it is necessary to pass the signals through an "antialiasing filter" before sampling them. This filter is usually set to attenuate all frequencies above 40 percent of the sampling frequency. For example, if we want information from a signal up to 10 kHz, then we should sample the signal at 25 kHz.

Once the signal is sampled and digitized, we have to decide how it will be processed to gain the information that we want. If all we want is an rms value, the processing is simple. Most other signatures start with a Fourier transform, but there are many other possibilities, such as wavelet analysis or Wigner-Ville distributions. The important point here is that one should not assume that any particular processing algorithm will be the magic bullet to detect a fault—the correlation between faults and signatures is a process that must be carefully done.

Figure 10.1   Block diagram of the major components of a diagnostic system. Some typical realizations of each of the steps or modules are indicated.

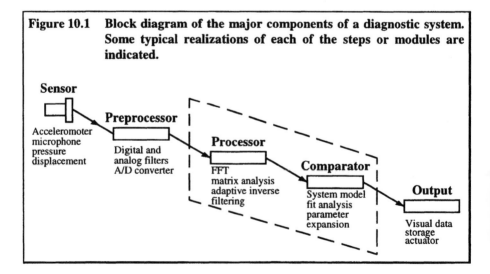

The signature can be any one of a set of values derived from the sensor signals. It may be the overall rms value of a signal, the magnitude of a line in the spectrum or some more complex signal, such as a modulation component or the cepstrum of the signal. The signature produces a "diagnostic signal," and the value of that signal is used in production QA to determine the condition of the product. The process of making a decision based on the value of the diagnostic signal is called *fault classification*.

There are two principal ways in which fault classification is carried out as sketched in Figure 10.2. One is to set a threshold for the diagnostic signal, and the fault is declared if the signal exceeds the threshold as shown in Figure 10.2a. In this case the diagnostic signal is derived directly from the processed sensor data. The second procedure is to have a model of the product dynamics and to apply the known inputs to the model in parallel to the product under test. Signatures extracted from both the model and the product are compared, and the model parameters are adjusted until there is a fit between the two as shown in Figure 10.2b.

Figure 10.2   **A model-based diagnostic system matches the system output to a model and then uses parameter values for the best matching model to determine the nature and magnitude of the fault. This is referred to as "running the model in the background."**

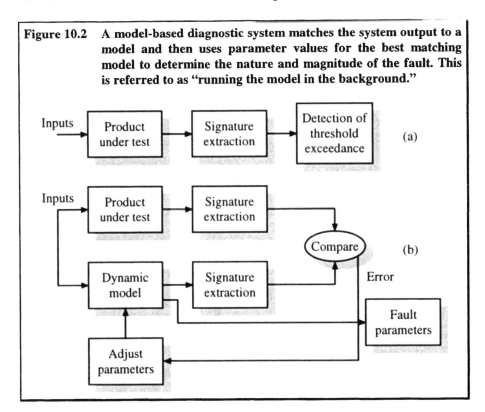

Model-based classification is a more complex procedure than signature thresholding, but it has the advantage of giving more direct information about the fault. However, it does require having a reliable dynamic model of the product that will predict abnormal behavior of the product as certain basic parameters are changed. Such models are relatively common for some products, such as turbomachinery, but do not generally exist for consumer products.

10.2
THRESHOLD SETTING AND AN EXAMPLE; PRINT HAMMERS

Whether it is an exceedance of a diagnostic signal derived from a sensor-based signature or an out-of-bounds parameter derived from model-based algorithm, a limit has to be set to make a decision that the product has failed the QA test. How that limit is set will be determined by explicit knowledge or intuitive ideas regarding the costs of a wrong decision.

Figure 10.3 shows the general idea. The amplitude of the diagnostic signal is the abscissa, and the probability density of recording the signal is the ordinate. It is assumed that a larger value of the signal is more likely associated with a failed condition, so the distribution of failed products has a higher mean than that for good products. But a particular value of signal might be achieved by either a good or a failed product, so choosing any particular value of the signal will inevitably result in some good products being labeled faulty, and faulty products being labeled good.

Figure 10.3   A classic Bayes' decision model tries to locate the decision level to incur least cost. An important but often unknown parameter is the cost of a failure of either or both types.

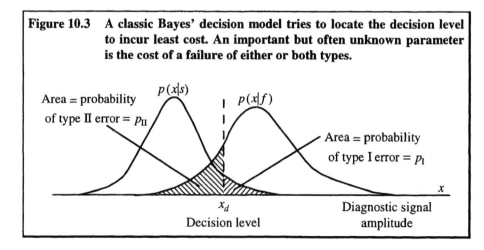

The probability of a good product being labeled good is called a *type I error*, and the probability of a good product being labeled bad is called a *type II error*. The areas labeled in the figure as p_I and p_{II} are the probabilities that these errors occur. If we set the decision level so these probabilities are equal, that is a "maximum likelihood decision" in that the choice we make is the more probable one. But that is not necessarily the decision we want to make.

Suppose we are operating a paper mill, there is an indication that a bearing is noisy on one of the production rolls, and we are one day away from a scheduled maintenance stop. The cost for stopping production to repair a bearing that is not really faulty is very high, but the cost of continuing to operate the roll with a failed bearing for another day is much less. In this case we are likely to set the threshold for declaring a faulty condition very high.

On the other hand, suppose we are producing braking systems for automobiles, and a pressure test on the brake line indicates the possibility of a leak in the system. In this case, we are likely to set the threshold for declaring a faulty condition very low because the cost of checking the system for a nonexistent leak is very small compared to installing a faulty system in a vehicle.

The theory of this process says that the minimum cost for both types of error combined is achieved if we set the decision threshold so that the expected cost of both types of error is equal,

$$p_I K_I = p_{II} K_{II} \qquad (10.1)$$

where K_I is the individual cost of a type I error and K_{II} is the individual cost of a type II error. Although neat from a theoretical point of view, it is rarely the case that one knows accurately either the forms of distribution shown in Figure 10.3, or the individual costs shown in Equation 10.1. Most of the time thresholds are set based on some intuitive notion of the costs and benefits involved, but the rationale presented does give a reason why acceptance thresholds are set to fail on either a very few or a larger number of products in QA testing.

Of course, if the two distributions are separated so that there is little overlap, then the decision of where to set the threshold is greatly simplified. An example is a case of print hammers made of sintered metal formed by inserting metal powder in a mold and compressing and heating it to form the part. It was discovered that some hammers would last indefinitely in service, whereas others, apparently built to the same specifications, would fail in a short time.

In studying a large number of the failures, the company determined that the resonance frequencies of the hammers were associated with the variations in life. As a test applicable to production, the procedure sketched in Figure 10.4 was devised. As hammers are produced they are allowed to fall on a hard surface and as they bounce off they radiate sound at their free resonance frequencies. This sound is picked up by a microphone and recorded.

A typical spectrum of the sound is shown in Figure 10.5. The hammer has two closely spaced resonances at approximately 14.5 and 15 kHz and another resonance at about 26.6 kHz. A histogram of the number of occurrences of the lower two frequencies in detail is shown in Figure 10.6, where we have distinguished the two lower resonance frequencies by ■ and □. We see that both resonances show a bimodal form, similar to that of Figure 10.3.

Figure 10.4 **The resonance frequencies of sintered powdered metal line printer hammered are measured by letting them fall onto a hard plate. As they bounce off they vibrate and radiate sound at their resonance frequencies.**

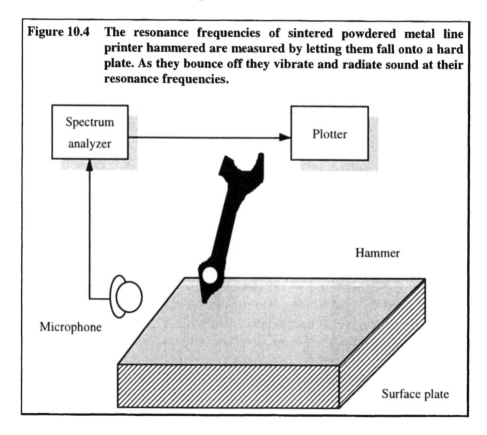

Figure 10.5   Spectrum of a typical acoustical pulse when the hammer is dropped against a surface plate.

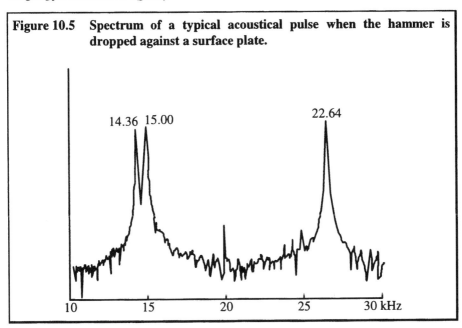

Figure 10.6   Histogram of the observed resonance frequencies for a group of hammers. The bimodal distribution indicates two classes of hammers in the population.

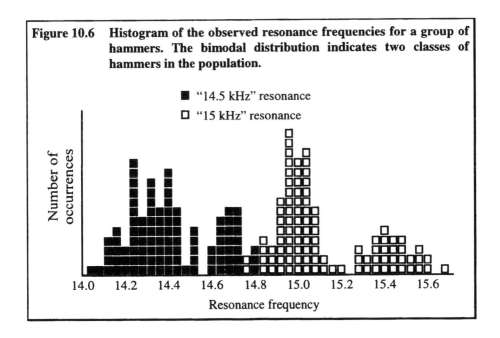

Because the amount of powdered metal injected into the mold is fixed, the resonance frequencies indicated the stiffness of the hammer as produced by sintering. It is perhaps surprising that it is the group of hammers at the higher frequency peak that come from a lot known to be defective, as we might expect the stiffer units to be better sintered and therefore more durable. But perhaps the sintering of that lot has been excessive, leading to a part that is more brittle. In any case, these resonance frequencies as the diagnostic signal, and the form of the distribution provides a direct indication of how the fault indicating threshold should be set.

10.3
THE "BATHTUB CURVE" AND VIBRATION-BASED DIAGNOSTICS

If one were to follow the life of a large number of produced items over their complete lifetimes and graph the rate of their failures, we might expect to produce a graph like that shown in Figure 10.7. There is a higher failure rate at earlier times, which then settles down to a fairly low rate for most of the expected life of the product. Then later, the failure rate increases again as the product ages. Diagnostic and prognostic systems will differ in their design and application depending on the part of the curve that one is concerned with.

Figure 10.7   **The classic "bathtub curve" showing the likelihood of failure during the life cycle of a product. Production line testing is used to reduce infant mortality while in-service diagnostics tries to catch midlife and old-age failures before they happen.**

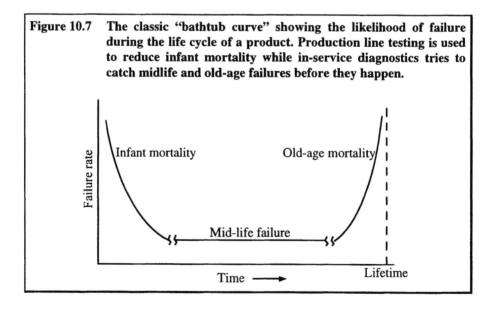

If the product is in service, diagnostic systems are concerned with incipient failures that will lead to a need for maintenance or replacement. Historical and empirical data are often used in such cases to guide in assessing the condition of the product. Figure 10.8 is an example of such data-based guidelines for rotating machinery, for which a great deal of experience has been gained to correlate vibration with machine condition. In this case, total vibration, typically measured on a bearing cap, is graphed with machine speed as the parameter. Because velocity, displacement, and acceleration are all graphed on the same chart (as we did in Figure 5.5), there is an assumption that the vibration occurs principally at the rotation frequency. The contours on the graph are essentially diagnostic signal thresholds for the different conditions cited.

Figure 10.8   Machinery fault analysis guides are provided by instrumentation companies to support users in making decisions about the health of machines during their service life.

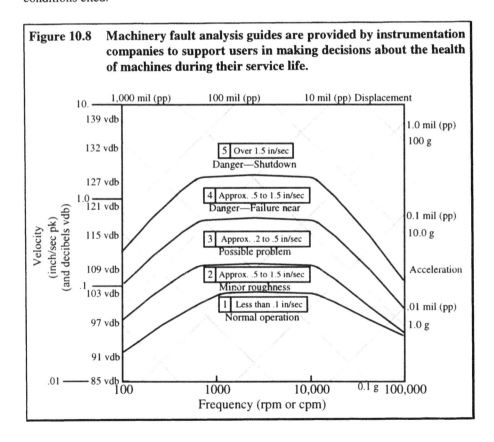

The principal purpose of production line QA testing is to catch those products that would fail early in the expected life of the product due to manufacturing defects. There are various forms that this testing can take. The reverberant noise booth example cited in Chapter 8 was a screening device to catch machines that exceeded a certain noise limit, but gave no diagnostic information regarding the cause of the excessive noise. Such a procedure requires that all units produced be tested. It is also possible of course to do 100 percent testing, which does provide diagnostic information. An example of this type of QA testing is given below.

Of course, companies would like to build products in such a way as to make testing on the line unnecessary. Testing in this case may take the form of an audit, sampling the product on the line and carrying out detailed analyses to detect any variations in assembly or components that would lead to quality problems. We shall provide an example of that situation also.

The two examples that we have discussed—the sewing machine noise booth and the print hammer resonance testing—both used sound as the basic signal to be analyzed. But measuring sound on the production line is inherently problematic because of interference from general plant noise and the space needed for an enclosure in which the measurement will be made. For that reason, vibration measurements are often preferred when one can correlate vibration with the defect(s) of concern.

If the defect is too much noise, as it was with the sewing machines, then the vibration has to be correlated with radiated sound, which may be difficult unless some sort of scan over the surface of the product is managed to estimate a ms velocity. In particular cases this has been accomplished by scanning a velocity sensing Laser Doppler pickup across the surface. If the defect is a noisy gear or the combustion pulse in a diesel engine, then a vibration measurement at a single location on the structure may be sufficient once the correlation between the defect and the vibration at that location has been established.

10.4

EXAMPLE OF 100 PERCENT TESTING FOR A GEAR DEFECT; GOLF CAR DRIVE SYSTEMS

A manufacturer of drive systems for golf cars had units returned from the customer because of a fluctuating tone of the gear mesh frequency as the golf car rolled to a stop. In this case, the cause of the offending sound was known, but the standard dyno test of the drive unit during production could not simulate the coast down load condition, and the spectrum analyzer being used could not track the dropping gear mesh frequency and detect the modulation of the mesh tone. A time-frequency spectrogram of the vibration measured during coast down of a defective unit is shown in Figure 10.9. The modulation of the gear mesh tone is plainly evident in this spectrogram.

The new on-line QA system developed is sketched in Figure 10.10. It consists of a computer workstation and a mechanical system to both drive and load the drive unit under test. The main sensor signal in this case is a vibration signal, but there is also a shaft encoder that gives continuous information about the rotational speed of the drive. The workstation has the following functions:

1. To control the sequence of steps in the test procedure by providing a checklist for the operator and indicating if a connection or other action has not been taken.
2. To make sure that the sensor(s) are connected and working.
3. To start and stop the test sequence by causing the drive motor to be energized and deenergized at the proper times.

Figure 10.9 **The modulation of gear mesh tones as the unit slows can be seen in the spectrogram and heard; the diagnostic system has to detect this modulation and assess its strength.**

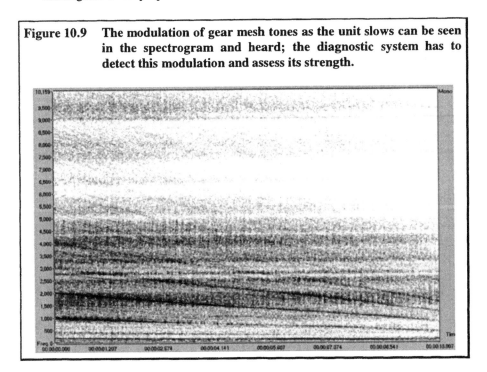

4. To receive data from the sensors and process it to detect the modulation shown in Figure 10.9 (we will indicate how modulation can be detected in the example that follows).

5. To determine whether the unit passes or fails, to indicate the result to the operator, and to archive the data into the plant network.

Figure 10.10 A production line quality assurance system to diagnose the modulation of gear tones combines the mechanical assembly for mounting the drive unit with a workstation that controls the test, analyzes the data, and communicates the results.

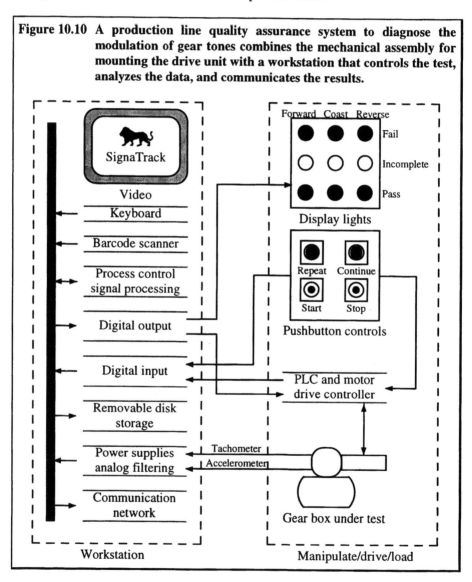

These functions are sketched out in Figure 10.10. We will discuss just how some of the signal processing to achieve some of the important signatures is carried out in the example that follows.

10.5
EXAMPLE OF A SAMPLING AND TEST SYSTEM FOR THE AUDIT LAB; WINDSHIELD WIPER MOTORS

When product is sampled for audit purposes, one is concerned primarily with controlling the process of manufacture because the theory is that good product is being produced, and one is looking for trends that may be precursors of a problem. Because various defects in a product will likely arise at different stages of production, if one is trying to control the *process*, then the diagnostics have to be able to separate the faults that will be introduced at those different stages. It is not good enough to say that a fault exists; the system must be able to classify the faults.

A windshield wiper drive consists of a DC permanent magnet motor with a rotor shaft that is knurled as a two-spiral worm as shown in Figure 10.11. The motor rotates at about 4000 rpm (67.5 Hz) producing a worm tooth engagement frequency of about 135 Hz. The worm engages a worm (helical) gear as the first stage of a two-stage reduction gear set that results in a final output shaft rotation frequency of about 1 Hz, which drives the wiper mechanism.

Figure 10.11   A windshield wiper motor drives a two stage reduction gear set through a two spiral worm machined into its shaft.

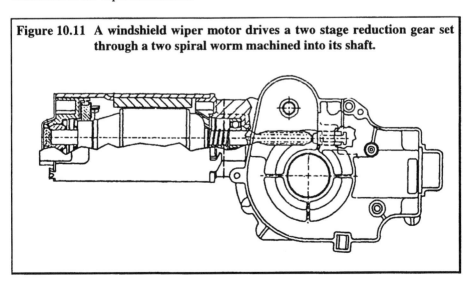

The audit lab system is sketched in Figure 10.12. The wiper motor is supported on an isolated mass by force gages, and there is also an accelerometer mounted to the reduction gear housing. The output shaft is loaded by a magnetic brake that can be voltage controlled for variable drag. The pulsating motor current (due to the motor commutator) is monitored for speed control. Speed of the motor is then controlled by the applied voltage, which in turn is controlled by the motor current pulsations. In this way, the motor speed and load can be independently controlled and programmed.

The sensor signals are motor current and voltage (the voltage fluctuates because the voltage supply has some internal resistance), the forces measured by the three-force gages, and the accelerometer output. These signals are shown in Figure 10.13. The once-per-rev periodicity in all the signals is apparent in all the signals, and the frequency spectrum of one of the force gages and the accelerometer are shown in logarithmic form (in dB) in Figure 10.14. The spectrum has a large number of frequency components at multiples of the motor rotation frequency of 67.5 Hz. This periodicity in frequency can be checked by taking a Fourier transform of the spectrum to create the power cepstrum, shown for the force gage signal in Figure 10.15a and b. The power cepstrum has peaks (rahmonics) at the reciprocal of the line spacing in the spectrum, which is about 15 msec that is, of course, the rotation period of the motor.

Figure 10.12 An audit lab test fixture for windshield wiper motors controls motor speed and load independently and processes electrical supply, vibration, and force signals.

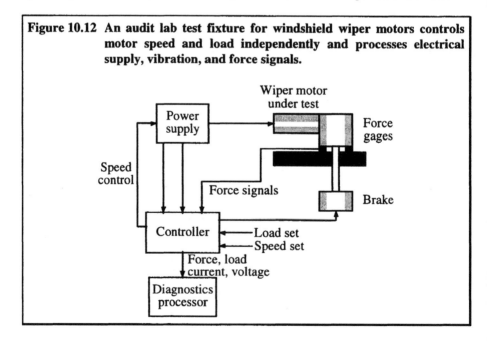

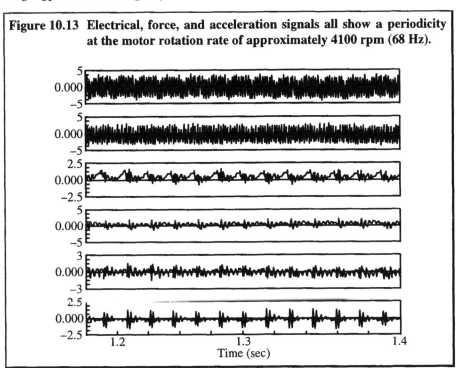

Figure 10.13 Electrical, force, and acceleration signals all show a periodicity at the motor rotation rate of approximately 4100 rpm (68 Hz).

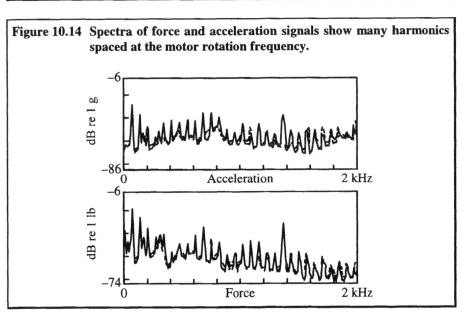

Figure 10.14 Spectra of force and acceleration signals show many harmonics spaced at the motor rotation frequency.

Figure 10.15   The power cepstra of force signals for

(a) a motor with a gouge on the commutator, and

(b) a normal commutator.

The first peak (rahmonic at 15 msec, the motor rotation period) of the cepstrum of the faulted motor is much larger than it is for the good motor.

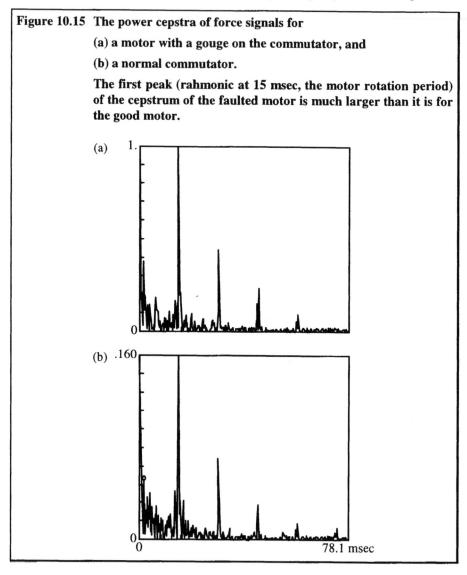

The first peak in the power cepstrum indicates the relative strength of tones in the spectrum to the random background. Figure 10.15 compares a force gage power cepstrum for a good motor compared to one for a motor with a gouge on the commutator. The first peak for the latter is substantially larger than the one for the former, indicating that the first cepstral peak may be a good diagnostic signal for commutator gouges.

If we now focus in on the peak in the force gage spectrum at 135 Hz, the worm mesh frequency, we get the result shown in Figure 10.16. These spectra, some for good motors and some for motors with a bent worms, all have similar appearance—a dominant peak at about 135 Hz and a large number of sidebands. Sidebands generally imply modulation. To explain the relation between the frequency spectrum and modulation, refer to Figure 10.17.

Figure 10.16 Narrow band spectra of force signals centered about the worm mesh frequency (2× motor rotation rate) for two good motors and two with bent worm shafts are not obviously different.

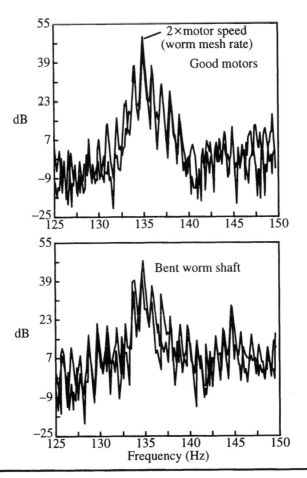

Figure 10.17 The amplitude and frequency modulation of a signal y_r can be determined from the real part of its Fourier transform and its Hilbert transform y_i, obtained from the imaginary part of the Fourier transform.

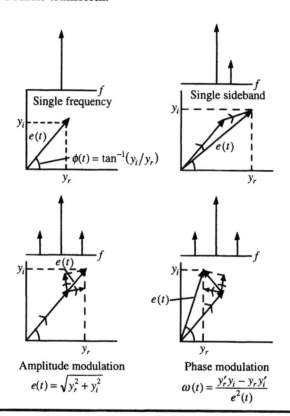

Consider a signal with the spectrum shown in the upper left corner, a single line at some frequency f_c. We can represent the signal as a phasor in the complex plane rotating at frequency f_c, where the projection along the horizontal (real) axis is the actual laboratory time waveform y_r and the projection along the vertical (imaginary) axis is its complex conjugate y_i. The amplitude of the signal is of course

$$e(t) = (y_r^2 + y_i^2)^{1/2} = A,$$ (10.2)

and the instantaneous phase of the signal is

$$\varphi(t) = \tan^{-1}(y_i/y_r) = 2\pi f_c\, t\,. \tag{10.3}$$

The instantaneous frequency of this signal is

$$\omega(t) = \frac{d\varphi(t)}{dt} = \frac{y_r \dot{y}_i - \dot{y}_r y_i}{A^2} = 2\pi f_c. \tag{10.4}$$

Now consider the next situation in the upper right corner of Figure 10.17 where another frequency component at frequency f_u has been added. The laboratory signal is still the real part of the summation of the two phasors representing these components, but now both the amplitude and the instantaneous frequency will fluctuate with time because as the original phasor rotates, the new component rotates even faster (as it has a higher frequency), and the sum of the two signals has a fluctuating amplitude and phase. We would describe this single sideband situation as a combination of amplitude and phase modulation, but note that Equations 10.2 and 10.3 still apply and will allow us to calculate the instantaneous amplitude and phase since we know y_r (our laboratory signal) and y_i can be computed from y_r if we have its complex Fourier amplitudes. The function $y_i(t)$ is called the Hilbert transform of $y_r(t)$.

Now consider the situation shown in the lower left corner of Figure 10.17 where a second sideband f_l has been added to the spectrum so that the sum of all three phasors simply increases and decreases the amplitude of the signal and the phase is unaffected. This situation represents pure amplitude modulation of the "carrier" at frequency f_c. But if we rotate the phases of the two sidebands as shown in the lower right of the figure, the spectrum of the signal does not change, but the modulation is now primarily phase modulation with a small amount of amplitude modulation. But in every case, Equations 10.2 and 10.3 will determine the instantaneous amplitude and frequency.

Now let's apply these ideas to the force signals whose spectra are represented in Figure 10.16. If we form the Hilbert transform of the force signals and compute the instantaneous amplitude for good motors, we get the patterns shown in Figure 10.18. The period of the signal covers two rotations of the output gear, and the variations in amplitude result from the varying load during a cycle because of the parking mechanism incorporated in the mechanism. There is a start-up rise in the signal and a transient at midpoint due to the processing algorithm, and it is not real.

Figure 10.19 shows the amplitude modulation when the worm has deformed threads. The amount of modulation is clearly greater when this fault is present. Figure 10.20 shows the modulation when the worm shaft is bent. In this case, the slow modulation of Figure 10.19 is present, but accompanied by a fast modulation that occurs at the worm rotation rate. The fast modulation tracks the fluctuating mesh load as the bent worm goes in and out of engagement. We are therefore able to distinguish between these two faults by the presence of the fast modulation, whereas the slow modulation indicates a worm fault in general.

Figure 10.18 The amplitude modulation of the worm mesh signal is fairly smooth for two good motors for two revolutions of the output gear (start-up transient due to signal processing).

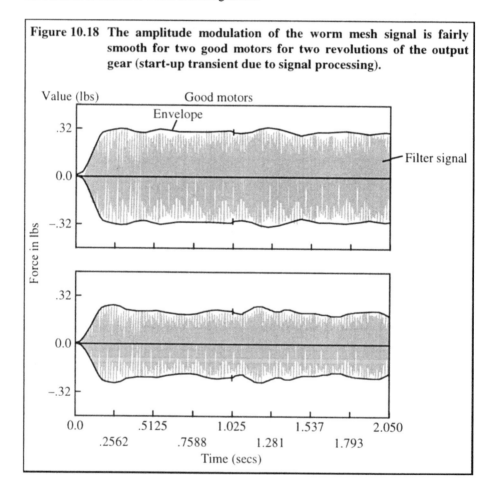

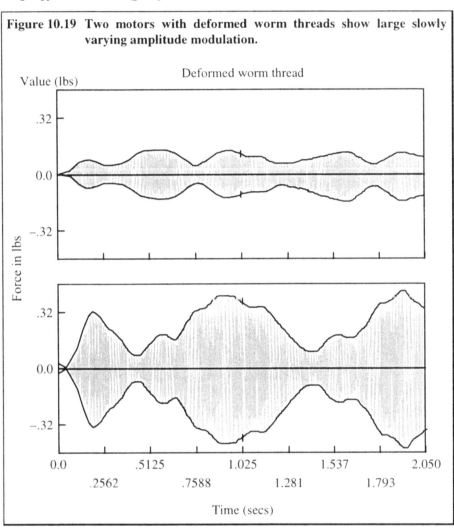

Figure 10.19 Two motors with deformed worm threads show large slowly varying amplitude modulation.

If we analyze the force signals of good motors for their phase modulation, which we will express in terms of instantaneous frequency, we get the signals shown in Figure 10.21, again with transients at the 1-second mark due only to processing. In this case, the average frequency is about 135 Hz, with a natural slow variation through the cycle, again due to the loading due to the parking mechanism. If we now look at the frequency modulation for motors with deformed worm threads shown in

Figure 10.22, we see a greater degree of fluctuation, although we can still follow the trend of average rotation speed or mesh engagement in the data. However, if we look at the pattern that results if the worm is bent in Figure 10.23, the rapid fluctuation in instantaneous frequency is so large that the average trend is more difficult to see.

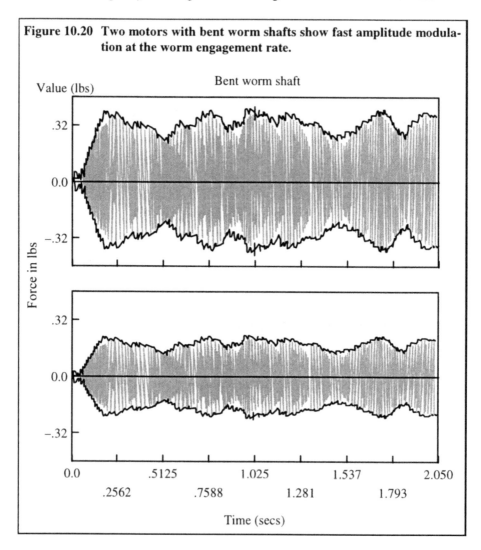

Figure 10.20 **Two motors with bent worm shafts show fast amplitude modulation at the worm engagement rate.**

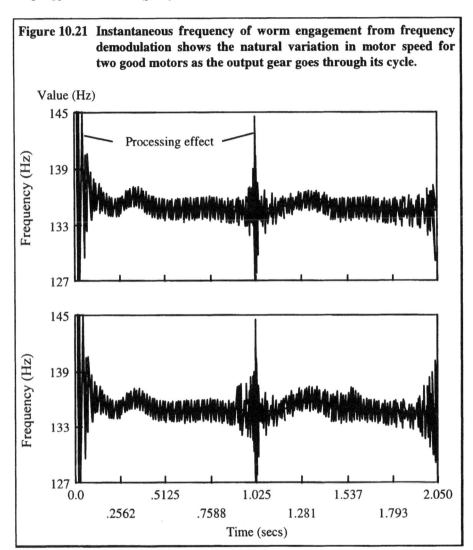

Figure 10.21 Instantaneous frequency of worm engagement from frequency demodulation shows the natural variation in motor speed for two good motors as the output gear goes through its cycle.

Figure 10.22 The instantaneous worm mesh frequency shows a slow frequency modulation for two motors with deformed worm threads.

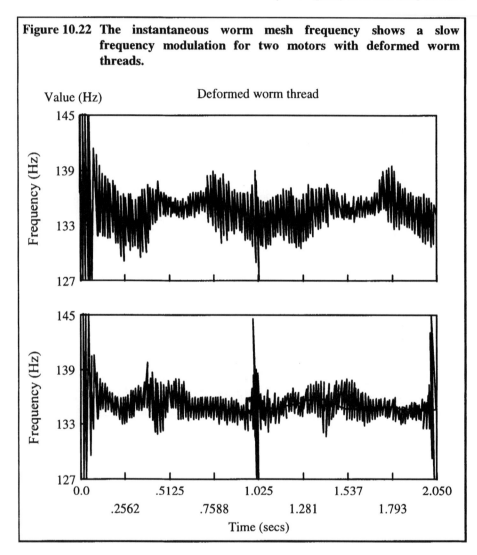

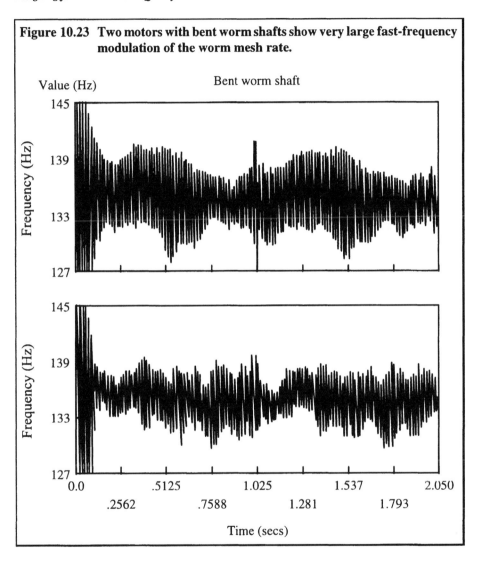

Figure 10.23 Two motors with bent worm shafts show very large fast-frequency modulation of the worm mesh rate.

These examples show us that different faults produce different effects in the signatures, but they do not indicate whether different faults create unique patterns in the diagnostic signals. Exhibit 10.1 presents the positive correlations between a set of faults and a set of signatures for this diagnostic system. We see from the pattern that the set of signatures (several of which we have not discussed) do reveal a unique pattern for each of the listed faults. We also see that the cepstrum (first peak) would be a good signal to detect motors with faults, because it is present for all faults, but the cepstrum is useless for the purpose of separating faults.

Exhibit 10.1
A "truth table" that correlates wiper motor faults with the diagnostic signals studied shows a unit pattern of signatures for each fault, but not a one-to-one correspondence between faults and signatures.

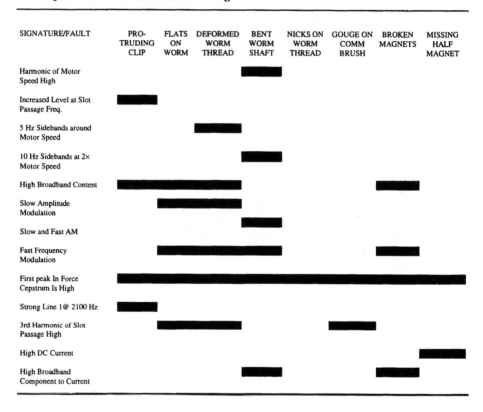

SIGNATURE/FAULT	PRO-TRUDING CLIP	FLATS ON WORM	DEFORMED WORM THREAD	BENT WORM SHAFT	NICKS ON WORM THREAD	GOUGE ON COMM BRUSH	BROKEN MAGNETS	MISSING HALF MAGNET
Harmonic of Motor Speed High				■				
Increased Level at Slot Passage Freq.	■							
5 Hz Sidebands around Motor Speed			■					
10 Hz Sidebands at 2× Motor Speed				■				
High Broadband Content	■	■	■	■			■	
Slow Amplitude Modulation		■	■	■				
Slow and Fast AM				■				
Fast Frequency Modulation	■	■	■	■			■	
First peak In Force Cepstrum Is High	■	■	■	■	■	■	■	■
Strong Line 1 @ 2100 Hz	■							
3rd Harmonic of Slot Passage High		■	■	■		■		
High DC Current								■
High Broadband Component to Current				■		■		

REFERENCES

1   Malen, D. E. and R. A. Scott, "Improving Automobile Door Closing Sound for Customer Preference", *Noise Control Engineering Journal,* **41**, #1, July–August 1993, pp 261–271, based on the PhD thesis of D. Malen: *Engineering for the Consumer: Methodology for Preliminary Design*, (Univ. of Michigan, Ann Arbor MI, 1992).

2   The Stevens method is described in *Noise and Vibration Control*, Ed. L. L. Beranek (Washington DC: Inst. of Noise Control Eng., 1988) pp 561–568. The Zwicker method is presented in E. Zwicker and H. Fastl, *Psychoacoustics, 2ᵈ ed.* (Berlin: Springer-Verlag, 1999) Section 8.7.3.

3   A thorough discussion of the use and evolution of NC Curves is given in *Noise and Vibration Control*, Ed. by L. L. Beranek (Washington DC: Inst.of Noise Control Eng., 1988).

4   Myers, R. H. and D. C. Montgomery, *Response Surface Methodology* (New York: John Wiley & Sons, 1995) pp 297 ff.

5   Many examples, one is Harris, C. *Handbook of Noise Control, 2d ed.*, (New York: McGraw-Hill Book Co., Inc., 1979).

6   Den Hartog, J. P., *Mechanical Vibrations, 4h ed.* (New York: McGraw-Hill Book Co., Inc., 1956).

7   Suh, I-S and R. H. Lyon, "An Investigation of the Sound Quality of I. C. Engines", (Warrendale, MI: Proc. 1999 Noise and Vibration Conference SAE, 1999) pp 481–488. based on the PhD thesis of In-Soo Suh "An Investigation of the Sound Quality of I. C. Engines", Dept of ME (MA: MIT, 1998).

8   *Electric Motor Handbook*, Ed. by H. W. Beaty and J. L. Kirtley; (New York: McGraw-Hill Book Co., Inc., 1998), Chapter 4.

9   Lyon, R. H., *Machinery Noise and Diagnostics* (Boston: Butterworth-Heinemann, 1987), Section 2.4.

10 Slack, J. W. and R. H. Lyon, "Piston Slap", contribution to Engine Noise, Ed. by R. Hickling and M. Kamal (New York: Plenum Press, 1982); based on the PhD Thesis of James Slack, "Piston Slap Noise in Diesel Engines", Dept. of ME (MA: MIT).

11 Lyon, R. H., *Machinery Noise and Diagnostics* (Boston: Butterworth-Heinemann, 1987), Section 3.12.

12 Lyon, R. H., *Machinery Noise and Diagnostics* (Boston: Butterworth-Heinemann, 1987), Section 4.3.

13 Lyon, R. H. and R. G. DeJong, *Theory and Application of Statistical Energy Analysis, 2d ed.* (Boston: Butterworth-Heinemann, 1995).

14 Unglenieks, R. J. and R. J. Bernhard, "Prediction and Verification of Energy Flow in a Structure using an Energy Finite Element Approach" (Warrendale, MI: Proc. Noise and Vibration Conference, SAE, 1995) pp 593–600.

15 *Mechanical Engineer's Handbook, 2ᵈ ed.*, Ed. by M. Kutz (New York: John Wiley & Sons, 1998) Section 18.5.5.

16 Parsons, N. E., R. G. DeJong and J. E. Manning, " Demonstration of Noise Control for the Cummins NTC-350 Heavy Duty Truck Diesel Engine," Final Contractor Report to EPA Office of Noise Abatement, Contract No. 68–01–4737, June 1982.

17 Lyon, R. H., *Machinery Noise and Diagnostics* (Boston: Butterworth-Heinemann, 1987), Section 5.6.

BIBLIOGRAPHY

(from basic to advanced)

Bergeijk, W., J. R. Pierce and E. E. David, *Waves and the Ear* (New York: Anchor Books, Doubleday Co.).

Rossing, T. D., *The Science of Sound, 2ᵈ ed.* (Reading, Mass.: Addison-Wesley, Longman, 1990).

Norton, M. P., *Fundamentals of Noise and Vibration for Engineers* (Cambridge: Cambridge University Press, 1989).

Gelfand, S. A., *Hearing, 3ᵈ ed.* (New York: Marcel Dekker, 1998).

Pierce, A. D., *Acoustics* (New York: Acoustical Society of America, 1994).

Lyon, R. H., *Machinery Noise and Diagnostics* (Boston: Butterworth-Heinemann, 1987).

INDEX